Getting Started With BeagleBone

Matt Richardson

SEBASTOPOL, CA

Getting Started With BeagleBone

by Matt Richardson

Published by Maker Media, Inc., 1005 Gravenstein Highway North, Sebastopol, CA 95472.

Maker Media books may be purchased for educational, business, or sales promotional use. Online editions are also available for most titles (*http://my.safaribooksonline.com*). For more information, contact O'Reilly Media's corporate/institutional sales department: 800-998-9938 or *corporate@oreilly.com*.

Editor: Brian Jepson
Production Editor: Christopher Hearse
Cover Designer: Jason Babler
Interior Designer: David Futato
Illustrator: Marc de Vinck

October 2013:   First Edition

Revision History for the First Edition:

2013-09-26:   First release

See *http://oreilly.com/catalog/errata.csp?isbn=9781449345372* for release details.

ISBN: 978-1-449-34537-2

[LSI]

Contents

Foreword

Matt is leading the charge to make technology serve you, the individual, and a new generation of innovators. BeagleBone Black is his newest, strongest—and most affordable—tool for building understanding, mastery, and just outright fun electronics projects. As one of the creators of this tool that enables just about anyone to sense, control and manage the data in the world around them, I'm obviously proud of what it can do. No amount of pride, however, is going to help you understand BeagleBone's capabilities or how to master them. Matt's contribution with this book is a piece previously missing from the Beagle-verse and one I'm confident will help you in your journey along the path he's paving.

When I was quite young, it was two books that set me on the path to understanding what could be accomplished with programmable electronics: *Getting Started in Electronics* by Forrest M. Mims III and *Getting Started with TRS-80 BASIC* by George Stewart. At the time, my experience with each programming and electronics was a separate endeavor. Programming was, at the time, the way you made use of a computer. The computer wasn't burdened with the storage of family photos or even precious business data, because my mom's business data was safely removed using floppy disks and stored away from my exploration. I was able to type in instructions to do whatever I could imagine, as long as I didn't open the box.

Far away from the computer, I was making runs to Radio Shack and buying components to build circuits that blinked LEDs and reacted to the ambient light in the room. It was almost a decade before I started connecting components up to microprocessors. As much as I had enjoyed modifying the games I'd typed into the computer such that I'd always win, having my programs interact with the physical world around me was an entirely new source of fire in my soul. All the everyday technology around me took on new meaning as I could understand how to make it myself and make it behave as I wanted.

When Gerald Coley, the hardware designer of BeagleBone Black and all of the *BeagleBoard.org* boards, approached me in 2007 to do something new with TI's ARM processors to bring the technology to a much wider audience, the idea of bringing back something closer to my childhood programming and electronics experiences fell naturally out of our discussions with our colleagues. Gerald's passion for excellence in electronics is something that has proven itself invaluable to the BeagleBoard.org community and me personally. Gerald has certainly never been one to be satisfied with typical notions of what is good enough.

With the emergence of so many new do-it-yourself electronics tools in recent years, I'm thrilled that many aspects of my childhood electronics experiences are once again available to technology-minded individuals looking to build something new of their own for the first time. It seems, however, that this split—between programmable computers that let you do what you expect to do with a computer, such as browse the web or even act as a web server, and devices that are great to talk to real-world components like motors, temperature sensors and light switches—is still quite prevalent. BeagleBone spans that divide.

Thanks now to Matt's effort with this book, I'm quite hopeful many more people will learn what programmable electronics can enable for them and experience what Gerald has offered to all of us. Even more, I hope it is a part of educating the next generation at any age how to make technology serve them, rather than merely living with someone else's idea of the perfect gadget who's purpose is to serve that someone else's goals.

—Jason Kridner
BeagleBoard.org cofounder and author/maintainer of BoneScript

Preface

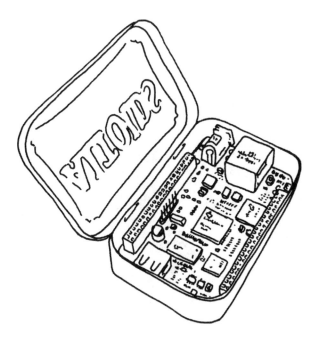

I first noticed the BeagleBone in 2011. At the time, most of my projects involved the Arduino, so I was quite curious when I saw a board that looked a little bit like an Arduino but acted more like a full computer. It seemed a bit complicated, so I was initially skeptical that I would be able to get *anything* working with it. Nonetheless, I ordered a BeagleBone and eagerly anticipated its arrival.

When it arrived, I was first amused by its dimensions. It fit in the palm of my hand and could even be enclosed within an Altoids tin. In fact, it fit almost too perfectly inside the tin. The radius of the rounded corners seemed to indicate that it was designed for such an enclosure. I'd later learn that, indeed it was designed that way.

After a lot of Internet searching and brushing up on how to write scripts within Linux, I had the BeagleBone blinking an LED, a common first step with hardware development platforms. Soon thereafter, I was reading the state of buttons, pulling images from a webcam, printing text with a receipt printer, and connecting the board to the Internet.

My first big project with the BeagleBone was called the Descriptive Camera. It worked a lot like a regular camera: point it at a scene that you want to capture and then hit the shutter button. But that's where the similarities with a camera end. Instead of saving a photograph, this prototype camera outputs a text description of the scene that you've captured. And it even spits it out of the front of the camera like a Polaroid print.

The Descriptive Camera didn't use any fancy computer vision algorithms to convert the image into text. It actually used crowd sourcing. After hitting the shutter button, the photo would be uploaded to Amazon's Mechanical Turk service, where you can pay people online to do small tasks like transcribing audio, identifying terms in a contract, or in this case, describe a photo. After the person submitted the text, it would be outputted by the camera's printer.

The BeagleBone was the perfect platform for this endeavor. Making a project that brings together a USB webcam, an Internet connection, buttons, LEDs, and the receipt printer, all while enclosing it in a small box would have been very difficult with many of the other platforms out there. As a tool, the BeagleBone is so capable and flexible that I could have created this same exact project in so many different ways.

But I know from experience that having a tool that's so versatile can make things hard when you're just starting out. There's no right way to do any single thing, so you can feel paralyzed before you've even begun.

My hope is that this book will get you through that initial phase. It will give you just enough of the basics in a few different realms so that you can start digging deeper on your own. Having a few different ways to do the same thing means you can settle on the way that you're most comfortable with and focus on making your vision a reality.

Conventions Used in This Book

The following typographical conventions are used in this book:

Italic

> Indicates new terms, URLs, email addresses, filenames, and file extensions.

`Constant width`

> Used for program listings, as well as within paragraphs to refer to program elements such as variable or function names, databases, data types, environment variables, statements, and keywords.

`Constant width bold`

> Shows commands or other text that should be typed literally by the user.

`Constant width italic`

> Shows text that should be replaced with user-supplied values or by values determined by context.

This icon signifies a tip, suggestion, or general note.

This icon indicates a warning or caution.

Using Code Examples

This book is here to help you get your job done. In general, you may use the code in this book in your programs and documentation. You do not need to contact us for permission unless you're reproducing a significant portion of the code. For example, writing a program that uses several chunks of code from this book does not require permission. Selling or distributing a CD-ROM of examples from MAKE books does require permission. Answering a question by citing this book and quoting example code does not require permission. Incorporating a significant amount of example code from this book into your product's documentation does require permission.

We appreciate, but do not require, attribution. An attribution usually includes the title, author, publisher, and ISBN. For example: "*Getting Started With BeagleBone* by Matt Richardson (Maker Media). Copyright 2014, 978-1-4493-4537-2."

If you feel your use of code examples falls outside fair use or the permission given here, feel free to contact us at *bookpermissions@makermedia.com*.

Safari® Books Online

 Safari Books Online is an on-demand digital library that lets you easily search over 7,500 technology and creative reference books and videos to find the answers you need quickly.

With a subscription, you can read any page and watch any video from our library online. Read books on your cell phone and mobile devices. Access new titles before they are available for print, get exclusive access to manuscripts in development, and post feedback for the authors. Copy and paste code samples, organize your favorites, download chapters, bookmark key sections, create notes, print out pages, and benefit from tons of other time-saving features.

Maker Media has uploaded this book to the Safari Books Online service. To have full digital access to this book and others on similar topics from MAKE and other publishers, sign up for free at http://my.safaribooksonline.com (*http://my.safaribooksonline.com/?portal=oreilly*).

How to Contact Us

Please address comments and questions concerning this book to the publisher:

MAKE
1005 Gravenstein Highway North
Sebastopol, CA 95472
800-998-9938 (in the United States or Canada)
707-829-0515 (international or local)
707-829-0104 (fax)

MAKE unites, inspires, informs, and entertains a growing community of resourceful people who undertake amazing projects in their backyards, basements, and garages. MAKE celebrates your right to tweak, hack, and bend any technology to your will. The MAKE audience continues to be a growing culture and community that believes in bettering ourselves, our environment, our educational system—our entire world. This is much more than an audience, it's a worldwide movement that Make is leading—we call it the Maker Movement.

For more information about MAKE, visit us online:

MAKE magazine: *http://makezine.com/magazine/*
Maker Faire: *http://makerfaire.com*
Makezine.com: *http://makezine.com*
Maker Shed: *http://makershed.com/*

We have a web page for this book, where we list errata, examples, and any additional information. You can access this page at:

http://oreil.ly/Getting-Started-BeagleBone

To comment or ask technical questions about this book, send email to:

bookquestions@oreilly.com

Acknowledgements

We'd like to thank a few people who have provided their knowledge, support, advice, and feedback to *Getting Started with BeagleBone*:

Brian Jepson
Marc de Vinck
Jason Kridner
Gerald Coley
Tom Igoe
Clay Shirky
John Schimmel
Phillip Torrone
Limor Fried
Justin Cooper
Andrew Rossi

1/Embedded Linux for Makers

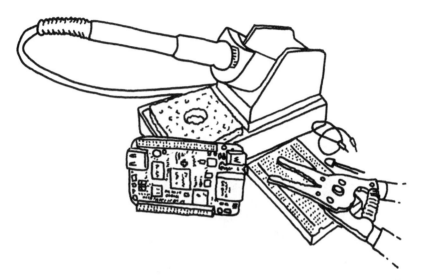

If you're familiar with Linux, you probably think of it firstly as a computer operating system, like OS X or Windows. It's usually running on user desktops and powering servers. But recently, Linux can also be found within many consumer electronics devices. Whether they're inside a cell phone, cable box, or exercise bike, *embedded Linux systems* blur the definition between computer and device.

This blurriness has made its way into the maker realm and that's great because it's putting more powerful tools in the hands of regular people, not just those who design electronics for a living.

Many makers who work with electronics love microcontroller platforms like the Arduino, but as the complexity increases in their projects, sometimes an 8-bit microcontroller doesn't have the power or capabilities to do what they need it to do. For example, if you want to use a camera and computer vision algorithms to detect dirty dishes in your sink, it might be a good idea to explore your options with embedded Linux development boards. These boards are generally more powerful and more capable than their 8-bit cousins and are sometimes the perfect solution for projects that are too complex for our beloved Arduino.

Not only that, but as the price of embedded Linux platforms drops, the community of support around them grows, which makes them much more accessible to novice and intermediate makers than ever before.

The BeagleBone (Figure 1-1) is an embedded Linux development board that's aimed at hackers and tinkerers. It's a smaller, more barebone version of their BeagleBoard. Both are open source hardware and use Texas Instruments' processors with an ARM Cortex-A series core, which are designed for low-power mobile devices.

Figure 1-1. *The Original BeagleBone*

Why Use BeagleBone?

These days, a typical microcontroller-based board costs around $20, while the BeagleBone Black retails for $45 at the time of press. Other than a more powerful processor, what are you getting for the extra money?

Built-in networking

Not only does the BeagleBone have an on-board Ethernet connection, but all the basic networking tools that come packaged with Linux are available. You can use services like FTP, Telnet, SSH, or even host your own web server on the board.

Remote access

Because of the built-in network services, it makes it much easier to access electronics projects remotely over the Internet. For example, if you have a data-logging project, you can download the saved data using an FTP client or you can even have your project email you data automatically. Remote access also allows you to log into the device to update the code.

Timekeeping

Without the need for extra hardware, the board can keep track of the date and time of day and can be updated by pinging Internet time servers using the network time protocol (NTP), ensuring that it's always accurate.

Filesystem

Just like our computers, embedded Linux platforms have a built-in filesystem, so storing, organizing, and retrieving data is a fairly trivial matter.

Use many different programming languages

You can write your custom code in almost any language you're most comfortable with: C, C++, Python, Perl, Ruby, or even a shell script.

Multitasking

Unlike a typical 8-bit microcontroller, embedded Linux platforms are capable of sharing the processor between concurrently running programs and tasks. This means that if your project needs to upload a large file to a server, it doesn't need to stop its other functions until the upload is over.

Linux software

Much of the Linux software that's already out there can be run on the BeagleBone. For example, when I needed to access a USB webcam for one of my projects, I simply downloaded and compiled an open source command line program which let me save webcam images as JPG files.

Linux support

There's no shortage of Linux support information out on the web and community help sites like Stack Overflow (*http://stackoverflow.com/*) come in handy when a challenge comes along.

What About Raspberry Pi?

There's a lot of buzz around Raspberry Pi, and while it's quite similar to the BeagleBone, there are certainly a few differences. For one, the Raspberry Pi is meant as a low-cost computer to encourage the younger generation to learn about how computers work and how to program them. Because of that, the hardware, software, and documentation are geared towards that objective. On the other hand, the BeagleBone is aimed more broadly at people interested in embedded Linux development boards and therefore has more options for connecting hardware and has a more powerful processor.

If you're interested in exploring the world of Raspberry Pi as well, I encourage you to check out *Getting Started with Raspberry Pi (http://shop.oreilly.com/ product/0636920023371.do)* (Maker Media), which I co-wrote with Shawn Wallace.

Intended Audience

Even though embedded Linux development boards are becoming easier to work with, it does take some skill (or at least patience and persistence) to use them if you're just starting out. This book assumes you know your way around a typical computer, be it with OS X, Windows, or Linux. While it's not necessary, it will help to know how to get around the Linux command line as well. This book will equip you with only the very basics of Linux skills so that you can work through the examples and projects. As you start to create your own projects with BeagleBone, a good foundation in Linux will be incredibly helpful. Luckily, there's an enormous community of support around Linux, so help is usually one web search away.

This book will also walk you through the fundamentals of programming the board with JavaScript and Python. There's a lot to learn in both languages so I can't cover all the details, but I'll point you to additional resources for learning more. If you prefer coding in another language, this book can still give you a good idea of how to do a few special BeagleBone tricks.

Feedback

I encourage you to contact me with any feedback as you read this book. I hope to be able to incorporate your suggestions into future editions. My email address is *mattr@makezine.com*. You can also find me on Twitter with the name @MattRichardson (*https://twitter.com/MattRichardson*).

2/The Basics and Getting Set Up

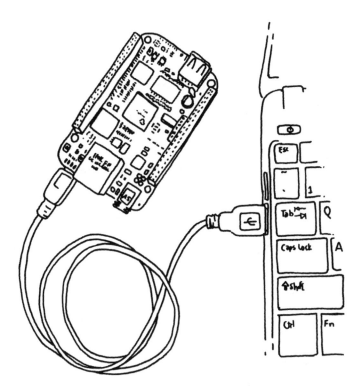

Your first major step into the world of Beagle-Bone is to hook it up and get to a command prompt so you can start working with files and executing commands. From there, you'll be able to customize the system to suit your preferences and start creating your own projects.

But before you connect anything, let's take a closer look at the BeagleBone. There are two versions of the board: the original BeagleBone and the newer BeagleBone Black. For most of this book, you'll be able to do everything with either board, with a few exceptions that I'll note.

It's quite easy to tell the boards apart. The original BeagleBone is mostly white with black lettering and the BeagleBone Black is mostly black with white lettering. The main improvements with the BeagleBone Black are a faster processor, more memory, on-board storage, and on-board video output. Not only that, but the BeagleBone Black is half the price of the original BeagleBone.

Tour of the Board

When you take a close look at the BeagleBone (Figure 2-1), you'll see that there are a lot of parts on it, some very small. Luckily, you don't have to understand what each part is in order to get the most out of the board. Here are a few of the more significant components:

1. *The Processor*. This is essentially the brains of the whole operation. Tasked with most of the heavy lifting, the processor on the original BeagleBone puts it in the same league with the iPhone 4 in terms of power. If you like to hear the numbers, it's a 720MHz ARM Cortex-A8 equipped with 256 MB of DDR2 RAM. If you have a BeagleBone Black, it'll be a 1GHz chip with 512MB of DDR3 RAM.

2. *The Power Connector*. Your BeagleBone needs 5 volts and 500 mA of direct current to operate. Most generic 5V DC power adapters with a 2.1mm barrel jack connector will power the board. It's important to know that even if a power connector will fit into this jack, it doesn't necessarily mean that it's providing 5 volts. Right nearby the jack is a small power a protection chip in case you accidentally provide over five and up to 12 volts. It will protect your board and won't let it power on if you connect too much voltage. Still, it's probably best to make sure you're plugging only 5 volts into the board.

3. *Ethernet Port*. This is a standard RJ45 Ethernet port, which will come in handy for Internet-connected projects. You can connect it directly to a router, or you can also share your computer's WiFi connection through Ethernet to the BeagleBone.

4. *Reset Button*. Press this button to reboot the board. Just like with your computer, it's best to trigger a reboot properly from within the operating system, otherwise file corruption could occur. This button might come in handy if your system locks up.

5. *USB Host Port*. Just like your computer, the BeagleBone is equipped with a USB port. This will let you attach a slew of hardware—including keyboards, mice, and Wi-Fi adapters—to your board.

6. *Onboard LEDs*. Next to the power connector, you have an LED to indicate when power is applied to the board. There are also four LEDs next to the reset button that can be programmed by you with software. By default, LED 0 will show a "heartbeat" when the system is running. LED 1 will blink when the MicroSD card is being accessed. LED 2 will blink when the CPU is active, and LED 3 will blink when the on-board flash memory is being accessed (on the BeagleBone Black).

7. *Expansion Headers*. These two expansion headers, labeled P8 and P9, allow you to integrate your BeagleBone into electronics projects. The pins can be configured for a number of different functions, which we'll dive into in Chapter 4.

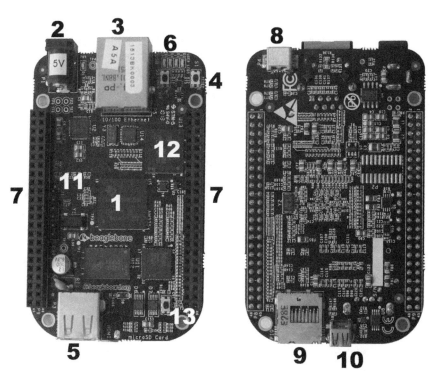

Figure 2-1. *The major components of the BeagleBone Black*

8. *Mini USB Port*. This USB port allows your BeagleBone to act as a device when you connect it to your computer. Your computer will not only provide power to the board over USB, but it also acts as a means of communicating with it. You can also access BeagleBone reference information stored on-board; it will simply appear as a storage device when you plug it in to your computer. If you choose to power your board through

this port, the processing speed will be reduced to decrease its power consumption.

9. *MicroSD Card Slot.* Unlike most computers, the BeagleBone doesn't have a hard drive and instead uses a MicroSD card to store the operating system, programs, and your data. On a BeagleBone Black, the operating system is stored on the onboard flash memory (see below) and can be updated using the MicroSD card slot.

10. *Micro HDMI Port (BeagleBone Black only).* To connect the BeagleBone Black to a monitor or television, use the Micro HDMI port. By looks, it's easily to confuse it with the Mini USB port, so if you have trouble plugging the cable in, check that you've got the right port.

11. *Serial Header (BeagleBone Black only).* While both the original Beagle-Bone and the BeagleBone Black have serial outputs for accessing the terminal, only the BeagleBone Black breaks out one of the serial ports in its own header. The layout on this header makes it easy to connect an FTDI TTL-232 cable or breakout board so that you can use the text based terminal via USB.

12. *On Board Flash Memory (BeagleBone Black only).* The BeagleBone Black sports on-board flash memory and can therefore be booted without a MicroSD card inserted. In the technical manuals, this memory is referred to as the *eMMC*.

13. *Boot Switch (BeagleBone Black only).* Holding down the boot switch when you power on your BeagleBone Black instructs the hardware to boot from the MicroSD card instead of the on-board flash memory.

What You Need

As you get familiar with the BeagleBone, you'll notice that there are a lot of different ways to get things done. Depending on what way works best for you and what your project calls for, you don't necessarily need everything on this list. However, having all of these components handy will help you try out the projects in this book:

- BeagleBone
- 5V DC power supply
- Ethernet cable
- USB A to mini B cable
- Solderless breadboard
- Jumper cables
- LEDs
- Resistors, assorted

- Buttons
- Switches
- 2K Potentiometer
- TMP35 or TMP36 temperature sensor
- Spare 4 GB MicroSD card
- MicroSD card reader

And if you have a BeagleBone Black, you might also want:

- HDMI capable monitor
- Micro HDMI to HDMI cable or adapter
- Keyboard
- Mouse
- USB hub
- 3.3V FTDI cable (see "Connecting via Serial over USB" on page 13 for more information)

The Operating System

Just like a computer, the BeagleBone has an operating system. By default, it uses Linux, which is free and open source. While there are many different flavors, or *distributions* of Linux out there, *BeagleBoard.org* offers a distribution called Ångström that's tailored for the board.

A factory-fresh BeagleBone Black will have Ångström preloaded onto the onboard flash memory, or eMMC. If you have an original BeagleBone, it will come with a microSD card with Ångström on it. Since development on this distribution happens rapidly, it's a good idea to stay up-to-date to the latest version. There was a large software update in April 2013. The examples throughout this book will expect that you have at least this version. Appendix A walks you through how to create an up-to-date MicroSD card.

While it's possible to use other distributions of Linux or even non-Linux operating systems on the BeagleBone, I recommend using the Ångström distribution since it's packaged up to install easily on the board and it's what the folks from *BeagleBoard.org* work with when they are testing and developing with the board.

Connecting to Your BeagleBone

As I mentioned before, there are a lot of ways to get things done with the BeagleBone and especially so when it comes to ways of connecting to its *command line terminal*. From the command line, you'll be able to build and execute programs, run administrative tasks, get information about your board, and much more.

Most of the time, I prefer having the BeagleBone connected to my home network router via Ethernet. This way, I can connect to its command line with *SSH* (Secure Shell), manage files via *SFTP* (SSH File Transfer Protocol), and have the BeagleBone access the Internet to download code and software packages when necessary.

When it comes to troubleshooting issues that cause an unstable network connection, I find connecting to the command line via serial over USB to be handy. Below I'll walk you through a few of the different ways that you can connect.

Connecting via USB and Installing Drivers

The BeagleBone itself is preloaded with documentation and drivers that will help you connect to it from your computer.

1. If you have an original BeagleBone, be sure that a MicroSD card with the latest version of the BeagleBone Ångström image is inserted into the slot.

2. Connect the BeagleBone to your computer via a USB A to mini-B cable

3. After about 20 seconds, a drive called BEAGLEBONE should appear in your filesystem's disk volume list. Open that drive and double click on the START HTML document (*START.htm*) to open it up in your default web browser.

4. Follow the instructions in the "Install Drivers" section on that page for your operating system.

5. In your web browser, go to *http://192.168.7.2/* to open Bone 101. This page is being served by your BeagleBone and has a lot of information about the board, including some interactive examples of Bonescript, a JavaScript library written for the BeagleBone.

Feel free to explore this if you're interested. We'll come back to using Bonescript in Chapter 7, but for now, let's get to a command prompt.

Connecting via SSH over USB

1. Open your terminal and connect to the BeagleBone:

   a. If you're on a Mac, open the Terminal application found in */Applications/Utilities/*. At the $ prompt, type `ssh root@192.168.7.2`

   b. If you're using Linux, type `ssh root@192.168.7.2` at your command-line prompt.

c. On a Windows PC, download and install PuTTY (*http://www.chiark.greenend.org.uk/~sgtatham/putty/download.html*). Enter 192.168.7.2 as the host address, making sure that "SSH" is selected and press connect. When it shows you the prompt "login as:" type root and press enter.

2. The first time you connect, you'll be warned about connecting to an unknown host. You can dismiss this message.

3. There's no password set by default, so if it prompts you, just hit enter.

4. You know you're connected when you see the prompt:

```
root@beaglebone:~#
```

Connecting via SSH over Ethernet

From time to time, you'll want to connect to your BeagleBone over the network instead of over USB.

1. If you have an original BeagleBone, be sure that the included MicroSD card is inserted into the slot.

2. Connect the BeagleBone to your router with an Ethernet cable and then plug in a 5V power supply to the BeagleBone.

 It's easy to accidentally eject the MicroSD card when you're applying force to the board to plug in the cables. To avoid this, hold the BeagleBone by its length-side edges as you plug in the cables (Figure 2-2).

3. Connect to it via SSH:

a. If you're on a Mac, open the Terminal application found in */Applications/Utilities/*. At the $ prompt, type ssh root@beaglebone.local

b. If you're using Linux, type ssh root@beaglebone.local at your command-line prompt.

c. On a Windows PC, download and install PuTTY (*http://www.chiark.greenend.org.uk/~sgtatham/putty/download.html*). Launch PuTTY, then enter "beaglebone.local" as the host address, making sure that "SSH" is selected and press connect. When it shows you the prompt "login as:" type root and press enter.

Figure 2-2. *Connecting the Ethernet cord to the BeagleBone*

No Connection?

If you're on Windows and the host name beaglebone.local does not work, you may need to download Bonjour Print Services for Windows (*http://support.apple.com/kb/dl999*). You can also use the IP address of the board instead. Find it by logging into your router and looking for "beaglebone" on the DHCP clients list.

---

4. The first time you connect, your SSH client may warn you that the host is unknown. It's OK to accept the host key and dismiss this message.

5. There's no password set by default, so if it prompts you, just hit enter.

6. You know you're connected when you see the prompt: `root@beagle bone:~#` (Figure 2-3).

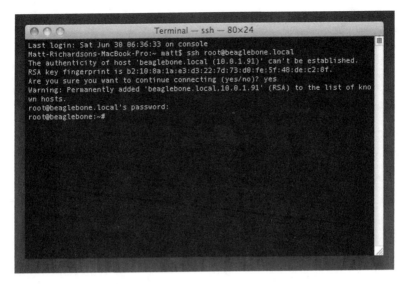

Figure 2-3. *SSH login screen*

Using a Keyboard, Monitor, and Mouse

If you have a BeagleBone Black, you can use it directly by connecting an HDMI monitor, keyboard, and mouse. Since there's only one USB host port on the BeagleBone, you'll need to use a USB hub to connect both the keyboard and mouse, unless of course your keyboard has a built-in hub. When you boot up the BeagleBone Black, you'll be presented with the GNOME desktop environment. To get to the terminal, click Applications, System Tools, then Terminal.

Throughout this book, we'll be doing a lot from the command line. Some operations, like creating, modifying and moving files around can also be done from the desktop environment if you prefer. Much of the desktop environment included on the BeagleBone is likely to feel familiar even if you're used to the Windows or Mac operating systems.

For more information on using the desktop environment, see Chapter 8.

Connecting via Serial over USB

You can also connect to your BeagleBone over USB via serial. This text-only connection is handy when you're experiencing networking problems. It also lets you see what's happening while the BeagleBone is booting up, before it has launched the services necessary to connect to it via the network. If you're able to connect via SSH, there's no need to do this now, but keep a mental note that this is an option should you run into issues logging in over the network.

The process on the BeagleBone Black is slightly different than on the original BeagleBone, so I've included separate instructions for each below. If you want to connect to the original BeagleBone via serial, you can use a basic USB A to Mini B cable (which you used in "Connecting via USB and Installing Drivers" on page 10). If you want to connect to the BeagleBone Black, you'll need to use a 3.3 volt FTDI TTL-232 cable or adapter board. These are much less common than USB A to Mini B cables, but can be purchased from vendors like Sparkfun (*https://www.sparkfun.com/products/9717*) and Adafruit (*http://www.adafruit.com/products/70*).

Connecting to the Original BeagleBone via Serial with OS X or Linux

1. Install the drivers in "Connecting via USB and Installing Drivers" on page 10 if you haven't already.

2. With the included MicroSD card inserted, connect the BeagleBone to your computer with a USB A to Mini B cable.

3. Open a terminal window and type:

   ```
   screen `ls /dev/{tty.usb*B,beaglebone-serial}` 115200
   ```

4. The screen will blank (besides a blinking cursor). Hit enter to display the login screen (Figure 2-4).

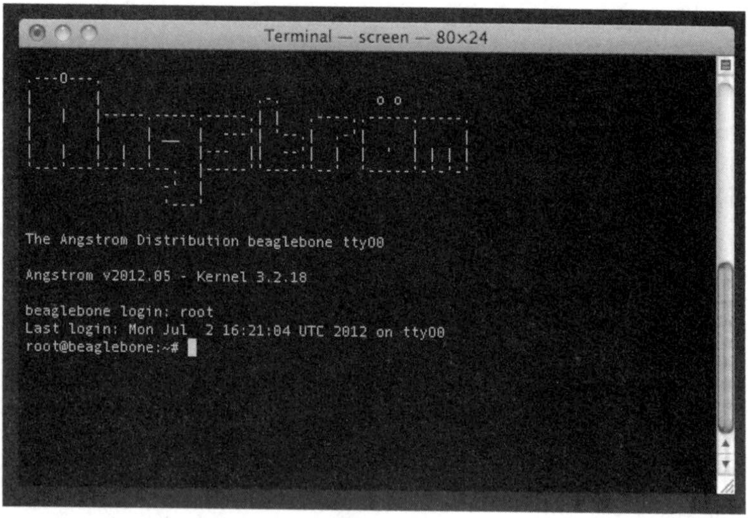

Figure 2-4. *The serial log in screen*

5. Log in with the user name root.

6. There's no password set by default, so if it prompts you, just hit enter.

7. To exit and disconnect from the BeagleBone, type Control-A and then K.

Connecting to the Original BeagleBone via Serial with Windows

1. If you haven't already, install the drivers in "Connecting via USB and In-stalling Drivers" on page 10.

2. With the included MicroSD card inserted, connect the BeagleBone to your computer with a USB A to Mini B cable.

3. Download and install PuTTY (*http://www.chiark.greenend.org.uk/~sgtatham/putty/download.html*). Launch PuTTY.

4. For the Connection Type, choose Serial.

5. Type in the name of the serial port for your connection. You might need to look in Device Manager (Windows Key+R, then type *devmgmt.msc*, click OK and look under Ports) to find it. On my system, it was COM7. Click OK.

6. For Speed, type 115200.

7. The rest of the defaults should be fine (see Figure 2-5). Click OK.

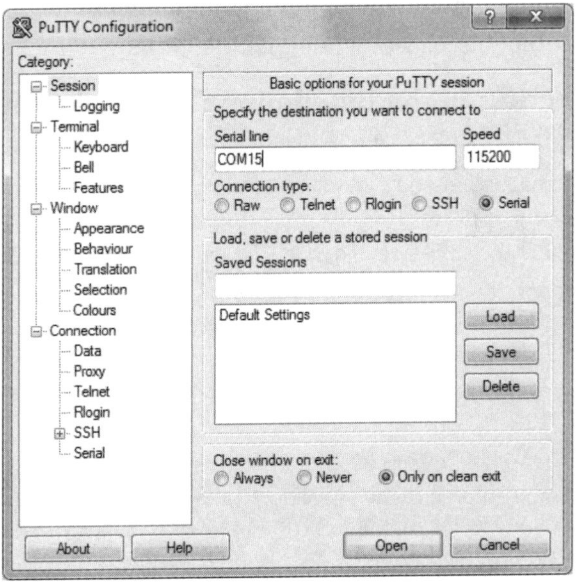

Figure 2-5. *PuTTY settings for Windows computers*

8. You'll see a flashing cursor in the terminal window. Press enter to be taken to the login screen.

9. Log in with the user name root.

10. There's no password set by default, so if it prompts you, just press enter.

Connecting to the BeagleBone Black via Serial with OS X or Linux

1. If you haven't already, install the drivers in "Connecting via USB and Installing Drivers" on page 10.

2. Connect the USB side of a 3.3V FTDI cable to your computer.

3. Connect the other side of the FTDI cable to the six male pins marked J1 on the BeagleBone Black. The FTDI cable's black wire should be closer to the "J1" label.

4. Open a terminal window and type:

```
screen `ls /dev/tty.usbserial-*` 115200
```

5. The screen will blank (besides a blinking cursor). Hit enter to display the login screen (Figure 2-4).

6. Log in with the user name root.

7. There's no password set by default, so if it prompts you, just hit enter.

8. To exit and disconnect from the BeagleBone, type CTRL+A and then K.

Connecting to the BeagleBone Black via Serial with Windows

1. If you haven't already, install the drivers in "Connecting via USB and Installing Drivers" on page 10.

2. Connect the USB side of a 3.3V FTDI cable to your computer.

3. Connect the other side of the FTDI cable to the six male pins marked J1 on the BeagleBone Black. The FTDI cable's black wire should be closer to the "J1" label.

4. Download and install PuTTY (*http://www.chiark.greenend.org.uk/~sgtatham/putty/download.html*). Launch PuTTY.

5. For the Connection Type, choose Serial.

6. Type in the name of the serial port for your connection. You might need to look in Device Manager (Windows Key-R, then type *devmgmt.msc*, click OK and look under Ports) to find it. On my system, it was COM7. Click OK.

7. For Speed, type 115200.

8. The rest of the defaults should be fine (see Figure 2-5). Click OK.

9. You'll see a flashing cursor in the terminal window. Press enter to be taken to the login screen.

10. Log in with the user name root.

11. There's no password set by default, so if it prompts you, just press enter.

Later in this book, I'll cover another way to connect to the BeagleBone to use the Cloud9 integrated development environment (IDE). For now, however, your best bet is to get your BeagleBone on your Ethernet network. This way, you can use SSH to get to the command line, use SFTP to manage files, and the BeagleBone will be able access the Internet, which will be required in Chapter 4.

3/Getting Around with Linux

```
root@beaglebone:~# cd /
root@beaglebone:/# ls
bin   dev  home  lost+found  mnt   run   sys  usr
boot  etc  lib   media       proc  sbin  tmp  var
root@beaglebone:/#
```

To the uninitiated, Linux may seem like a strange beast. It comes with a lot of power, customizability, and is heavily influenced by operating systems that date back to the early days of computing. Besides the price (free), the best part of using a Linux OS is the huge community of users, who have contributed their knowledge to the code itself and to helping other users.

While there are many different *distributions*, or flavors, of Linux, the people at BeagleBoard.org provide their version of the Ångström distribution of Linux for use with the BeagleBone. It's the software that's preloaded on the BeagleBone Black and is also available for download from their servers. Throughout this book, we'll be using Ångström, but it's possible to use other distributions of Linux and even other operating systems on the BeagleBone.

The Command Line

The main objective of Chapter 2 was to connect to your BeagleBone and get to a command line prompt. At this prompt, you can enter commands to start programs, work with files (such as create, delete, copy, and move them), compile your own programs, update system settings, and much more. By default, the BeagleBone's command line prompt will look like this:

```
root@beaglebone:~#
```

Let's take a close look at the prompt to see what each part means.

root

> This indicates the user you're logged in as, in this case, root. The user account root is considered the *superuser*, or administrator of the system. As root, you'll have unfettered access to most of the system's functions. But with great power comes great responsibility: it also makes it much easier to make changes that can make your board unusable.

beaglebone

> This indicates the hostname. This is how other computers on your network refer to your BeagleBone. Later in this chapter, you'll see how to change this in case you need something more descriptive like "toaster."

~

> This indicates the current *working directory*. It's your current location in the filesystem. If you run a command to create a file without specifying another location, the file will be created in the working directory. The tilde is shorthand for the logged in user's home directory. When you're logged in as root on the BeagleBone, the tilde indicates the location /home/root.

#

> This is the prompt for input. It also indicates that we're logged in as a superuser. If you were logged in as a regular user, the prompt for input would appear as a $.

Filesystem

Just like in many other operating systems, the Linux filesystem is an organized structure of files within folders, or *directories*. The root of the filesystem (not to be confused with the root user) is indicated by a forward slash (/). Within the root of the filesystem, there are a few main directories, most of which are listed in Table 3-1.

Table 3-1. *Directories in the root of the filesystem.*

bin	Programs and commands for users
boot	Files needed at boot time
dev	Files that represent devices on your system
etc	Configuration files
home	User home directories
lib	System libraries and drivers
media	Location of removable media such as USB flash drives and microSD cards
proc	Files that represent information about your system
sbin	System maintenance programs
sys	Files for accessing the BeagleBone's hardware

tmp	Temporary files
usr	Programs available for all users
var	System log files

Let's start using the command line now to explore the Linux filesystem on the BeagleBone. The first command you'll learn is pwd, which stands for *print working directory*. It tells you where you are in the filesystem.

```
root@beaglebone:~# pwd
/home/root
```

This indicates that you're in the directory named *root*, which is contained in the *home* directory, which is in the root of the filesystem. This particular location is the root user's home folder. In most cases, this is where you'll put projects that you work on as the root user.

Changing Directories

Let's say you want to change the working directory to the root of the filesystem. You can use the cd command:

```
root@beaglebone:~# cd /
root@beaglebone:/#
```

Did you notice that subtle change in the command line prompt? Instead of a tilde, there's now a forward slash, which indicates we've changed from the home directory to the root of the filesystem. To be sure, we can always execute pwd again:

```
root@beaglebone:/# pwd
/
```

 Because the tilde is shorthand for your home directory, you'll only see it when you're in your home directory or in one of its subdirectories. Otherwise, you'll see what's known as the *absolute path*, which starts with /, the root of the filesystem.

To change to the current directory's parent, use cd ..:

```
root@beaglebone:/# cd ~
root@beaglebone:~# pwd
/home/root
root@beaglebone:~# cd ..
root@beaglebone:/home# pwd
/home
root@beaglebone:/home# cd ..
```

```
root@beaglebone:/# pwd
/
```

The .. notation can also be used when typing out paths. For instance, if you were in the hypothetical path */home/root/myProject/sound* shown in the directory structure in Figure 3-1, you could change to */home/root/myProject/sound* by typing cd ../code.

Figure 3-1. *An example directory structure. You can change from /myProject/sound to /myProject/code by typing cd ../code*

Listing the Contents of Directories

And now that you're in the root of the filesystem, list the contents of the current working directory, with the command ls:

```
root@beaglebone:/# ls
bin   dev  home  lost+found  mnt   run   sys  usr
boot  etc  lib   media       proc  sbin  tmp  var
```

Now you can see all the directories explained in Table 3-1 (plus a few others). If you're interested in even more information about the contents of the current working directory, you can add the -l option to the ls command to display them in long format:

```
root@beaglebone:/# ls -l
total 56
drwxr-xr-x   2 root  root    4096 Mar 18  2013 bin
drwx------   2 xuser xuser   4096 Mar 18  2013 boot
drwxr-xr-x  13 root  root    3960 Jan  1 00:00 dev
drwxr-xr-x  65 root  root    4096 Jan  1 00:00 etc
drwxr-sr-x   4 root  root    4096 Mar 18  2013 home
drwxr-xr-x   9 xuser xuser   4096 Mar 18  2013 lib
drwx------   2 root  root   16384 Mar 18  2013 lost+found
drwxr-xr-x  11 root  root    4096 Jan  1 00:00 media
drwxr-xr-x   2 root  root    4096 Mar 18  2013 mnt
dr-xr-xr-x 106 root  root       0 Jan  1  1970 proc
drwxr-xr-x   7 root  root     160 Jan  1 00:00 run
drwxr-xr-x   2 root  root    4096 Mar 18  2013 sbin
dr-xr-xr-x  12 root  root       0 Jan  1 00:00 sys
drwxrwxrwt  11 root  root     240 Jan  1 00:00 tmp
drwxr-xr-x  12 root  root    4096 Mar 18  2013 usr
drwxr-xr-x  13 root  root    4096 Mar 18  2013 var
```

This listing gives a more complete picture, showing the permissions, owner, size, and date modified for each file or directory. Next, you're going to create some files and directories, so let's go back into the root account's home directory:

```
root@beaglebone:/# cd /home/root
root@beaglebone:~#
```

More Shortcuts Home

In the preceding example, you typed out the path to change the working directory to the root account's home directory, but there are a couple shortcuts that can take you there much more quickly. The tilde always refers to the logged in user's home directory, so typing cd ~ will also return you to your home directory. And if that's too much to type, entering cd alone on the command line will also take you home.

Use ls to show the contents of your home directory:

```
root@beaglebone:~# ls
Desktop
```

As you can see, there's already a directory within your home directory called *Desktop*. You don't need to worry about it for now, but in case you're wondering, files that appear on the desktop in the graphical user interface are stored there. For more information, see Chapter 8.

Creating Files and Directories

Use the command mkdir to create a new directory:

```
root@beaglebone:~# mkdir myProject
```

Change to the newly created directory:

```
root@beaglebone:~# cd myProject
root@beaglebone:~/myProject#
```

If you want to create a new file and store some text in it, you can do all that with one command:

```
root@beaglebone:~/myProject# echo 'Hello, world!' > hello.txt
```

The job of the command echo is to output to the terminal whatever text follows the command. We use the greater than sign to redirect that output to the file *hello.txt*, which will be created if it doesn't already exist.

If *hello.txt* already exists, you'll overwrite it, so be careful using the redirection symbol.

Now if you list the contents of *myProject*, you'll see your new file in the listing:

```
root@beaglebone:~/myProject# ls
hello.txt
```

To see the contents of the file, you can use the command `cat`:

```
root@beaglebone:~/myProject# cat hello.txt
Hello, world!
```

If you want to append text to the end of a file, the command you type will look like this:

```
root@beaglebone:~/myProject# echo 'What a beautiful day!' >> hello.txt
```

Be sure you use two greater than signs, which means to *append* the output from `echo` to the end of the file. To see the results, use cat again:

```
root@beaglebone:~/myProject# cat hello.txt
Hello, world!
What a beautiful day!
```

Although we're using `cat` to display the contents of the file, `cat`'s main function is to concatenate files. In other words, it will take a series of files and appends one after the other. To try that out, first make a few file:

```
root@beaglebone:~/myProject# echo 'See you real soon!' > bye.txt
```

Now let's use `cat` to join *hello.txt* to *bye.txt* to create a new file called *greetings.txt*:

```
root@beaglebone:~/myProject# cat hello.txt bye.txt > greetings.txt
```

Now when you output the content of *greetings.txt*, you'll see the content copied from the first two files:

```
root@beaglebone:~/myProject# cat greetings.txt
Hello, world!
What a beautiful day!
See you real soon!
```

All these command line tools can make it fairly easy to work with files, but sometimes you might just want to get inside a file to view and edit its contents. For that, I recommend the text editor nano. To open your file in nano, simply type nano followed by the file you want to open.

```
root@beaglebone:~/myProject# nano greetings.txt
```

Within nano, you can use your arrow keys to move the cursor around to delete and insert text. To save, type Control-O and to exit, type Control-X. Nano can do a lot of stuff including searching, clipboard operations, and spell check. To view the help information within nano, type Control-G.

Of course, you can also create new files with nano. Just type the name of the file you want to create after the nano command and that file will be written when you save within nano.

Copying, Moving, and Renaming Files

Copying and moving files around from the command line is fairly straight-forward. To try it out, create a new directory called *archive* inside *~/myProject*:

```
root@beaglebone:~/myProject# mkdir archive
```

To copy a file into the *archive* directory, use the command cp followed by the name of the file you want to copy and then its destination:

```
root@beaglebone:~/myProject# cp hello.txt archive
```

Now try moving a file with the mv command. It's used the same way as cp, except of course it moves the file instead of copying it.

```
root@beaglebone:~/myProject# mv bye.txt archive
```

You can also use mv to rename a file but keep it in the same spot. Here's how to rename *greetings.txt* to *salutations.txt*:

```
root@beaglebone:~/myProject# mv greetings.txt salutations.txt
```

Deleting Files and Directories

Use the command rm to delete files.

```
root@beaglebone:~/myProject# rm salutations.txt
```

If you want to delete an entire directory and its contents, use rm with the option -r. Be careful when using this since there's no way to undelete!

```
root@beaglebone:~/myProject# rm -r archive
```

Setup

Just like with any computer, there are settings on the BeagleBone that you may want to tinker with to suit your preferences. You're not required to do any of this, but it may come in handy for some projects.

Date and Time

Most computers have a *real time clock*, a piece of hardware that keeps track of the date and time. RTC's frequently have a battery backup so that they can keep time even when there's no power supplied to the computer. Unfortunately, the BeagleBone does not have a real time clock. This means that when you boot up the BeagleBone, it will have no idea what the date and time are. To prove it, try the date command:

```
root@beaglebone:~# date
Sat Jan  1 00:28:06 UTC 2000
```

However, the BeagleBone can keep track of time as long as it's powered on. The time just needs to be set every time it boots up. Of course, it's possible to set the time manually, but this wouldn't make much sense if you need to do this every time you power on the board. Fortunately, we're able to set up the BeagleBone so that it can get the correct date and time through the Internet using *Network Time Protocol*, or NTP.

Set the Timezone

The first thing you'll need to do is set your timezone. There's a built-in database of timezones preloaded onto the board and you'll need make a link to the appropriate timezone file from */etc/localtime*. First, backup the old *localtime* file by renaming it *localtime.old*:

```
root@beaglebone:~# mv /etc/localtime /etc/localtime.old
```

Change to the directory */usr/share/zoneinfo* and list the contents to see the available timezone files, keeping in mind, it might be contained within one of the continent directories such as America. I'll be using */usr/share/zoneinfo/America/New_York*:

```
root@beaglebone:~# cd /usr/share/zoneinfo/
root@beaglebone:/usr/share/zoneinfo# ls
Africa     CST6CDT  Europe  GMT0               MST7MDT  Pacific  Universal
zone.tab
America     EET     GB      Greenwich  NZ       ROC      W-SU
Asia        EST     GMT     HST        NZ-CHAT  ROK      WET
Australia   EST5EDT GMT+0   MET        PRC      UCT      Zulu
CET         Etc     GMT-0   MST        PST8PDT  UTC      iso3166.tab
root@beaglebone:/usr/share/zoneinfo# cd America/
root@beaglebone:/usr/share/zoneinfo/America# ls
Anchorage  Caracas  Chicago  Denver  Los_Angeles  New_York  Sao_Paulo
```

You'll then create a *symbolic link* from the */etc/localtime* file to your desired timezone file. A symbolic link is like a forwarding notice for a file. You can place one in a directory and most software that wants to read or write the file will be redirected to the file in another location.

 The shell prompt in the following example is truncated slightly to fit on the screen without wrapping the text.

```
# ln -s /usr/share/zoneinfo/America/New_York /etc/localtime
```

Set the NTP Server

The next step is to set up an NTP server, which will reply with the time. Edit the ntpdate configuration file with nano:

```
root@beaglebone:/usr/share/zoneinfo/America# nano /etc/default/ntpdate
```

Update the line with NTPSERVERS="" so that it looks like this:

```
NTPSERVERS="pool.ntp.org"
```

 pool.ntp.org will redirect your to one of about 4,000 public Network Time Protocol servers. Feel free to pick your own NTP server if you wish.

Save the file by typing Control-O and exit nano with Control-X. Now run the ntpdate-sync command to have the BeagleBone fetch the latest time. If everything works, you won't see any feedback from ntpdate-sync. After a few moments, try executing the date command again:

```
root@beaglebone:/usr/share/zoneinfo/America# ntpdate-sync
root@beaglebone:/usr/share/zoneinfo/America# date
Tue Jul  2 22:35:11 EDT 2013
```

The date and time will now be tracked as long as your BeagleBone is powered on. The BeagleBone is also pre-configured by default to synchronize with the NTP server once an hour, so you may have a period of time after booting when the time is incorrect. You can always run ntpdate-sync to force it to synchronize immediately. See Appendix B for how to set up ntpdate-sync to happen automatically on start up.

Software Installation, Updates

If your BeagleBone has a connection to the Internet, you can easily install or upgrade software and code libraries. The software comes in *packages* and

Ångström uses the utility `opkg` to manage the process of installing and updating packages.

`opkg` keeps a list of available package versions within the BeagleBone's filesystem, so it's important to update this list before attempting to install or update software. To update the list of available packages, simply execute:

```
root@beaglebone:~# opkg update
```

You'll see some text scroll by as it downloads all the latest listings. If you want to update the software, the `opkg` command `upgrade` will upgrade all of your installed packages if there are new versions available. Depending on how many software packages have been upgraded since the release of the software you're running, this process may take a *very* long time (a few hours, even):

```
root@beaglebone:~# opkg upgrade
```

If there's a particualr piece of software or code library that you want to install, you can also install it by name using `opkg`. To install the version control system git, for instance, you would run:

```
root@beaglebone:~# opkg install git
```

The same is true of upgrading. For instance if you only wanted to upgrade Python, you could run:

```
root@beaglebone:~# opkg upgrade python
```

Changing the Hostname

By default, your BeagleBone identifies itself with the hostname `beaglebone` on your network so that you can access it via the web with `http://beagle bone.local` or SSH with `ssh root@beaglebone.local`. In most cases, you won't need to change this, but if you start to use many BeagleBones on the same network, you'll want to distinguish one from another.

To edit the hostname, execute the following, keeping in mind that the hostname can only contain letters, numbers, or a hyphen. It cannot contain any other characters such as $, #, or a space:

```
root@beaglebone:~# echo 'your-hostname' > /etc/hostname
```

So that your BeagleBone recognizes its own hostname as itself, you'll also want to edit the hosts file as well:

```
root@beaglebone:~# nano /etc/hosts
```

Change the line that says `127.0.0.1 beaglebone`, replacing beaglebone with your hostname. After you save the file and quit nano, you need to reboot your BeagleBone. After you log in again, you should see your custom hostname in the prompt.

Setting a Password

By default, there's no password set for the user root in the Ångström images provided by BeagleBoard.org. If you exclusively use the BeagleBone via USB or on your own private network, you don't need to set one, but it's probably a good idea to set a unique password if you use the BeagleBone on a network that you share with others.

To set or change the password, use the command passwd:

```
root@xively-box:~# passwd
Enter new UNIX password: [hidden]
Retype new UNIX password: [hidden]
passwd: password updated successfully
```

Shutting Down

Disconnecting the power or resetting the BeagleBone without properly shutting it down can corrupt data and cause problems with your operating system. To shut down the BeagleBone properly, issue the following command:

```
root@beaglebone:~# shutdown
```

You'll know when it's safe to disconnect the power when the user LEDs stop blinking and the power LED next to the DC barrel jack connector turns off. The board will turn back on when you reconnect it to power. If you just want to reboot the board, use the reboot command.

```
root@beaglebone:~# reboot
```

4/First Steps with Digital Electronics

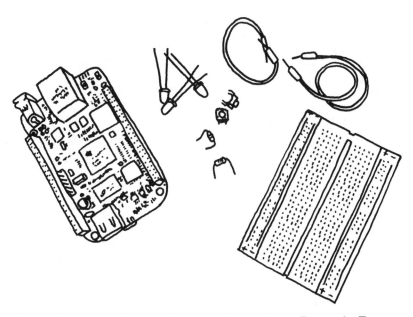

Development platforms like the BeagleBone provide an environment for a wonderful hybrid of both software and hardware hacking. In this chapter, you're going to test drive the *General Purpose Input/Output* pins (GPIO) on the board to get a basic sense of how they work and how you can read from and write to them.

No doubt you've noticed the two sets of headers that run along the edges of the board. They're labeled P8 (Figure 4-1) and P9. Each header has 46 pins and if you look closely you can see that pins 1, 2, 45, and 46 are labeled on each header (Figure 4-2). To identify pin numbers in the middle, you'll have to count off from the pins on the end.

Figure 4-1. *Header P8*

Figure 4-2. *Pin labeling on header P9*

The pins have many different possible functions, from controlling LCD screens, reading sensors, communicating with other electronics, and much more. Most pins can even be switched between modes to accommodate different possible functions. See Getting Things All Muxed Up for more information.

In this chapter, we're going to use a few of the pins in their *GPIO mode*. GPIO pins have two states: high and low. When a pin is high, that means it's connected to 3.3 volts. When a pin is low, it means that it's connected to ground. It's important to remember that digital pins must be either high or low. Reading the state of a pin that's not connected to anything will return unpredictable results. We say such a pin is *floating*.

 3.3 volts is the specified logic level of the BeagleBone. Other platforms, like the Arduino, may use 5 volts. Use only 3.3 volt logic components with the BeagleBone. Otherwise you can permanently damage the board.

Before we get into writing any software, let's take a look at how to do some basic digital pin control from the Linux command line. Once you get to know how the Linux kernel uses a *virtual filesystem* to read and write pins, it makes programming the BeagleBone much easier. It also means that you can then use any programming language you're comfortable with to read and write to the pins. As long as you can read and write files, you can work with GPIO.

To walk through the tutorials this chapter, you'll need the following components in addition to your BeagleBone, its power supply, and your computer:

- Solderless breadboard
- Jumper wires
- LEDs
- Resistors: 1×100 ohm, 1×10K ohm
- Momentary pushbutton or toggle switch

Connect an LED

A great way to get to know a new platform is simply getting an LED to blink, so let's start by wiring up an LED:

1. Shut down the BeagleBone (see "Shutting Down" on page 29). It's always good idea to shutdown the board and remove power before you start wiring things up to the pins.

2. Using a jumper wire, connect the negative rail of the breadboard to one of the BeagleBone's ground pins, which are located on pin one and two of header P8 and P9.

3. Using another jumper wire, connect the breadboard's positive rail to the BeagleBone's 3.3 volt source which are pins 3 or 4 on header P9.

 Be very careful not to accidentally connect the rail to the 5 volt source on pins 5 and 6. The GPIO pins can only handle 3.3 volts.

4. Place an LED in the breadboard so that the cathode side (the shorter wire) is connected to the negative rail and the anode side is in one of the rows of the breadboard.

5. Using a 100 ohm resistor, connect the anode side of the LED to another row on the breadboard.

6. Connect the other side of the 100 ohm resistor to pin 12 on header P8. On a 100 ohm resistor, the color bands printed on it will be brown, black, brown, and then gold or silver. Your circuit should look like Figure 4-3.

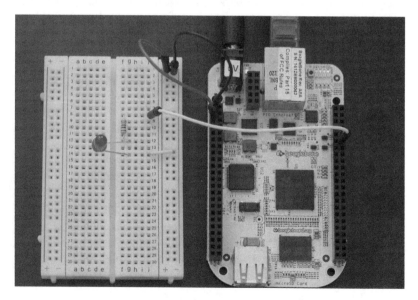

Figure 4-3. *Wiring up an LED to GPIO to pin 12 on header P8 (GPIO 44)*

Linux Signal Names

Unfortunately, the way that you'll refer to each pin within the Linux filesystem is not the same as the pin numbers printed on the board. Even if you look at BeagleBoard.org's System Reference Manual to get the signal name, there's still a little work done to determine the Linux signal name that you need. For the tutorials in this chapter, I've provided a quick reference to a few of the pins in Figure 4-4, but in case you want to use any of the other pins on the board, here's how to derive the number you'll need:

1. Download the BeagleBone's System Reference Manual from *http://beagleboard.org/bone*.

2. In the section of the System Reference Manual that shows the pinouts for P8, you can see that the default signal name for hardware pin 12 is "GPIO1_12". The signal names take the format of "GPIO[chip]_[pin]".

3. To determine the pin number that you'll use within Linux, multiply the chip number by 32 and add the pin number. So for signal GPIO1_12, you'd refer to it as GPIO signal 44 within Linux. (32 × 1 + 12 = 44).

 ## Getting Things All Muxed Up

You may have noticed that we've only labeled a few pins in Figure 4-4 and here's why: many pins can be assigned different functions, not just digital input and output. This feature is known as pin multiplexing or "pin muxing" and it can make things a little tricky. For this tutorial, I'm using pins that default to GPIO mode when the BeagleBone is powered on. Many of the pins default to other modes. Keep in mind that the defaults can also change as updated versions of Ångström are released for the board.

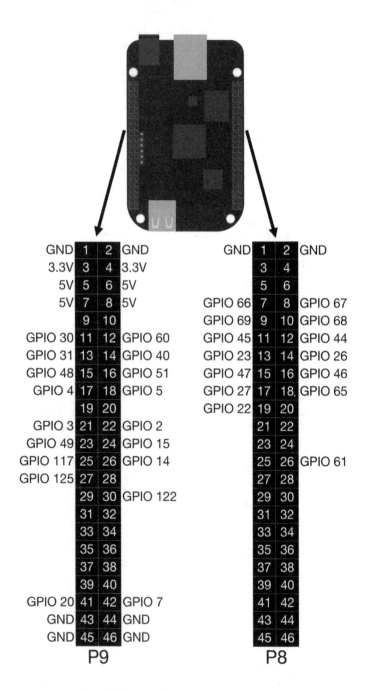

Figure 4-4. *The GPIO pins (BeagleBone illustration courtesy of the Adafruit Fritzing library)*

Output

Referring to Figure 4-4, you know to use GPIO signal 44 within Linux for pin 12 on header P8 and you know that it's set to work in GPIO mode by default. With that knowledge, you're ready to use the command line to control that pin:

1. On the command line, change to the gpio directory:

   ```
   root@beaglebone:~# cd /sys/class/gpio
   ```

   > When you're typing out long paths on the command line, there's a feature called command-line completion that will save you a lot of time. Start typing the first few letters of the directory or file and then hit tab key. If the system has a single file or directory that matches that, it will fill in the rest of the name for you. If there's no file or directory or there are multiple items that start with the same letters, you'll get a beep.

2. When you list the contents of the directory with the command ls you'll notice there's no folder for gpio signal 44. That's because we need to export the pin to userspace so that we can control it. To do that, use the echo command to write the number 44 to the export file:

   ```
   root@beaglebone:/sys/class/gpio# echo 44 > export
   ```

   > These pins can be used by many different possible functions and you don't want them to conflict with each other. For example, if you attach a BeagleBone expansion board, or *cape*, it may request access to a few of these pins for itself. Exporting the pin to the userspace is a way of saying to the Linux kernel, "I, the user, want to use this pin." The kernel will then warn you if it's already in use. If it's not in use, it will create the appropriate directory so that you can control the pin and it will then warn others that you have control of it.

3. Now when you type ls, you'll see the newly created **gpio44** directory.

   ```
   root@beaglebone:/sys/class/gpio# ls
   export  gpio44  gpiochip0  gpiochip32  gpiochip64  gpiochip96
   unexport
   ```

4. Change to that directory:

```
root@beaglebone:/sys/class/gpio# cd gpio44
```

5. Since you want to control an LED, you need to set the pin as an output by writing the word "out" to pin 44's `direction` file:

```
root@beaglebone:/sys/class/gpio/gpio44# echo out > direction
```

6. Now we're ready to set the pin high to illuminate the LED. Write 1 to the `value` file:

```
root@beaglebone:/sys/class/gpio/gpio44# echo 1 > value
```

7. We can then set the pin low and turn off the LED by writing a 0 to the `value` file:

```
root@beaglebone:/sys/class/gpio/gpio44# echo 0 > value
```

If you saw the LED illuminate after writing 1 to its `value` file and it turned off when you wrote 0 to the `value` file, congratulations! Feel free to experiment with the other pins in Figure 4-4.

Input

If you can control an output pin by writing to the `value` file, it stands to reason that you can read an input pin by *reading* the `value` file. By doing this, you can check the state of physical buttons and switches. Let's try it out now:

1. Place a push button into the breadboard so that it straddles the center channel. If you don't have a button handy, you can also use a toggle switch.
2. Connect one lead of the button to the positive rail.
3. Connect the other lead of the button to the input pin 11 on header P8. (See Figure 4-5 of BeagleBone wired up.)
4. Connect a 10K *pulldown resistor* from the ground rail to the button lead that connects to the input pin.

Remember that GPIO pins must be either high or low. That is to say, connected to either 3.3 volts or ground. The 10K pull-down resistor in the step above ensures that when the button is not pressed and the connection between 3.3 volts and the input pin is broken, the input pin is then connected to ground through the 10K resistor. Using a resistor ensures that when the switch is closed, the 3.3 volts (which is inclined to follow the path of least resistance) doesn't go directly to ground, creating a short circuit. Instead, it goes to the input pin.

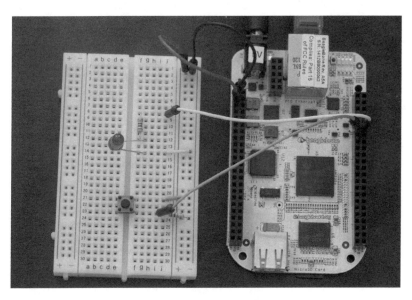

Figure 4-5. *A button and LED wired up to the BeagleBone*

5. Let's get back to the command line now. First, export the pin to the userspace and change to its directory. According to Figure 4-4, pin 11 on header P8 is GPIO signal 45, so that's what we'll export.

```
root@beaglebone:/sys/class/gpio/gpio44# cd ..
root@beaglebone:/sys/class/gpio# echo 45 > export
root@beaglebone:/sys/class/gpio# cd gpio45
```

6. Set the pin direction as an input:

```
root@beaglebone:/sys/class/gpio/gpio45# echo in > direction
```

7. Now, instead of writing the `value` file, we'll read it using the command `cat`:

```
root@beaglebone:/sys/class/gpio/gpio45# cat value
0
```

8. This should return 0 for low. This means that the pin is connected to ground. Now press and hold the button while you execute the `cat value` command again. If you have the button wired up correctly, you should now see a 1, indicating the pin is high (connected to 3.3 volts).

> An easy way to execute a command again is to hit the up arrow key on your keyboard and then hit enter. You can keep hitting up to scroll back through your history of commands. Just hit enter when you get to the command you want.

9. When you're done with the pins, be sure to unexport them from the userspace:

```
root@beaglebone:/sys/class/gpio/gpio45# echo 44 > /sys/class/gpio/unex
port
root@beaglebone:/sys/class/gpio/gpio45# echo 45 > /sys/class/gpio/unex
port
```

If you've successfully blinked an LED and read a button from the Linux command line, congratulations! It may seem trivial, but these examples represent the very basic foundation of digital electronics with the BeagleBone. And you're not limited to only LEDs, buttons, and switches. With the right circuitry, these could easily become tilt sensors, blenders, door security sensors, buzzers, motors, and much more.

Project: Networked Outlet Timer

Now that you know how to read and write pins from the command line, you can bring some of Linux's powerful features into the physical world. This project uses Linux's job scheduler, `cron`, and a PowerSwitch Tail relay to let you control a lamp or other A/C device on a set schedule. And since the BeagleBone is already network-enabled, it lets you change the settings of your outlet timer from the comfort of our computer, or even from the other side of the globe.

This project will also demonstrate the basics of using shell scripting as one method to programmatically execute commands.

If you haven't picked up a PowerSwitch Tail yet, you can test this project out using the LED you've already got connected to your BeagleBone. If you wired the LED up as directed earlier in this chapter, you won't even need to rewire anything.

Parts

Here's what you'll need to try this project out:

- BealgeBone
- 5V DC power supply
- Ethernet cable
- PowerSwitch Tail II
- Hookup wire
- Lamp

Wire up the Circuit

1. Using the hookup wire, connect pin 1 from the PowerSwitch Tail to pin 12 on header P8 on the BeagleBone.

2. Using another strand of hookup wire, connect pin 2 from the Power-Switch Tail to one of the ground pins on the BeagleBone. On both headers, they're on pins 1 and 2.

3. Plug the PowerSwitch Tail into a power source and plug your lamp into the other end of the PowerSwitch Tail.

4. Be sure the lamp is switched on. It won't light up yet since the Power-Switch tail is currently interrupting the power between the lamp and your power source.

The PowerSwitch Tail is a high voltage relay circuit that has been packaged up for easy use. It lets you use the 3.3 volt logic level signals that come from the BeagleBone to flip a switch to connect the A/C power to the device that's connected to it. So writing pin 12 to high will close the switch between the A/C wall outlet and our lamp.

Test the Circuit

1. Execute the following commands to actuate the relay and turn the lamp on. They should look familiar since they're the same commands we used to light the LED earlier in this chapter.

   ```
   root@beaglebone:~# echo 44 > /sys/class/gpio/export
   root@beaglebone:~# echo out > /sys/class/gpio/gpio44/direction
   root@beaglebone:~# echo 1 > /sys/class/gpio/gpio44/value
   ```

2. If the lamp turns on, you know you've got the circuit right. Otherwise, check that the lamp is switched on, that you've got the PowerSwitch Tail wired up correctly, and that you typed the commands correctly.

    If you received the error "write error: Device or resource busy", then you probably forgot to unexport pin 44 as directed at the end of "Input" on page 38. Make sure you run the unexport commands shown there before you try this example.

3. Now unexport the pin.

   ```
   root@beaglebone:~# echo 44 > /sys/class/gpio/unexport
   ```

Create the Shell Scripts

You can use a *shell script* to execute batches of commands. While shell scripts have the potential to be powerful and complex, they can also be very basic and as easy to write as entering the commands onto the command line.

1. Use cd to change to your home directory (remember that the shorthand for the home directory is ~) and create a new file called lightOn.sh for the shell script:

   ```
   root@beaglebone:~# cd ~
   root@beaglebone:~# nano lightOn.sh
   ```

2. This will launch the nano text editor. Type in the following code:

   ```
   #!/bin/bash ❶
   echo 44 > /sys/class/gpio/export ❷
   echo out > /sys/class/gpio/gpio44/direction ❸
   echo 1 > /sys/class/gpio/gpio44/value ❹
   ```

   ❶   This line is required for all shell scripts

❷   Export pin 44

❸   Set the direction to output

❹   Set the pin high

 For editing files while in a text-based environment there are plenty of options like vi, emacs, and pico. At risk of getting into a debate about which text editor is best, I'll say that I prefer using nano, but feel free to use whatever text editor you'd like. If you're new to this environment, start off with nano, which is arguably the easiest editor to use.

---

3. Type Control-X and type "y" to save the file when it prompts you.

4. Use nano to create another file called `lightOff.sh` with the following code:

```
#!/bin/bash
echo 0 > /sys/class/gpio/gpio44/value ❶
echo 44 > /sys/class/gpio/unexport ❷
```

❶   Set the pin low

❷   Unexport the pin

5. Type Control-X and type "y" to save the file when it prompts you.

6. To make both of those scripts executable, execute the following commands:

```
root@beaglebone:~# chmod +x lightOn.sh
root@beaglebone:~# chmod +x lightOff.sh
```

7. Now when you type `./lightOn.sh`, the light should turn on. When you type `./lightOff.sh`, the light should turn off. (You will need to be in the home directory to do this.)

Scheduling the Scripts

We're going to use Linux's built-in scheduler, `cron`, to set the light to turn on at 7pm and turn off at 4am.

1. On the command line, type the following command to edit your crontab (cron settings table) within nano (the `EDITOR=nano` part of the command

forces `crontab` to use the nano editor instead of the system default, which is usually vi):

```
root@beaglebone:~# EDITOR=nano crontab -e
```

2. Add the following lines to the end of the file:

```
0 19 * * * /home/root/lightOn.sh
0 4 * * * /home/root/lightOff.sh
```

3. Type Control-X and type "y" to save the file when it prompts you.

4. If the current time is between 7pm and 4am, you'll need to execute the *lightOn.sh* manually so that *lightOff.sh* doesn't return an error when it's executed at 4am.

A Crash Course in Cron

The formatting for the crontab might seem a little cryptic at first, but it's not as confusing as it seems. The cron scheduler will allow you to execute commands as frequently as once a minute or you can even set a command to execute many years in the future. Each crontab entry is on its own line with 5 space-delimited settings (sometimes 6, in the case of specifying years) followed by another space and then the command that should be executed as you can see in Table 4-1.

Table 4-1. *Cron entry for turning light on*

0	19	*	*	*	/home/root/lightOn.sh
Minute (: 00)	Hour (7pm)	Every Day	Every Month	Every Day of Week	path to command

Let's say you want a command to be executed every five minutes. See Table 4-2 for what the entry would look like.

Table 4-2. *Scheduling something for every five minutes*

*/5	*	*	*	*	/home/root/blinkLED.sh
Every Five Minutes	Every Hour	Every Day	Every Month	Every Day of Week	path to command

And if you want a command to be executed twice a week, on Monday and Thursday, see Table 4-3 for what the entry would look like.

Table 4-3. *Sheduling something for monday and thursday at 8am*

0	8	*	*	MON,THU	/home/root/ takeOutTrash.sh
Minute :00	Hour (8am)	Every Day	Every Month	Monday and Thursday	path to command

To execute a command on a particular year, you can add another space-delimited setting after the day of the week to specify the year (or years).

If you want to adjust the timing of the lamp, all you have to do is log into your BeagleBone and edit your crontab. You could even set up your router so that it's accessible from outside your home network, just be sure to set a password! In Chapter 6, we'll dig into Internet-connected projects even more.

5/Python Pin Control

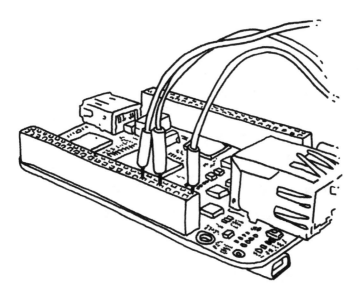

One of the strengths of embedded Linux platforms like the BeagleBone is being able to
choose the programming language that suits
you best. In many other cases, choosing a
hardware platform means you're also commit-
ting a particular language and development
environment. That's far from the truth with the
BeagleBone. The downside to the flexibility
that the BeagleBone affords is that there are
many different ways to do the same thing. This
chapter will walk you through one way to use
code to interact with the pins on the board.

When you write code for a microcontroller like the Arduino, most of the time, you write the code on your computer, compile it, and then upload the compiled binary to the microcontroller. However, on an embedded Linux system like the BeagleBone, the compiler or interpreter exists on the board itself. This chapter will show you how to use the Python interpreter on the Beagle-Bone to program the pins to behave the way you want.

Python (Figure 5-1) is a powerful programming language that's still very accessible for beginners. The code is relatively easy to understand and there are many libraries that help with complex tasks. For example, you'll use Adafruit's BeagleBone IO Python Library (*https://github.com/adafruit/adafruit-beaglebone-io-python*) to help with reading and writing the GPIO pins.

If Python is completely unfamiliar to you, don't worry. You'll learn how to use it for controlling pins as you work through the examples in this chapter. I'll also point to outside resources for learning more about Python if you're interested in digging a little deeper into the language.

Figure 5-1. *The Python Logo*

Installing Adafruit's BeagleBone IO Python Library

Out of the box, the BeagleBone doesn't have much support for using Python to access the input and output pins on the board. Just like we did in Chapter 4, it's possible to write Python code to manually read and write the GPIO files in the Linux virtual filesystem. However, this would take a long time. Luckily for us, the fine engineers at Adafruit have created an open source Python library for accessing the pins. I especially like this library because it follows a lot of the same conventions as the popular RPi.GPIO Python library that comes with the Raspberry Pi. This means that if you've made a project in Python for the Raspberry Pi, it may not be very difficult to port it over to BeagleBone.

Before you do anything else, you'll need to install some Python tools using opkg. From the command line, update the package listing. As mentioned in Chapter 3, it's a good idea to do this every time you update or install software:

```
root@beaglebone:~# opkg update
```

Then install the packages python-pip, python-setuptools, and python-smbus:

```
root@beaglebone:~# opkg install python-pip python-setuptools python-smbus
```

Now that you have the Python package management tool, pip, you can use it to install Adafruit's library:

```
root@beaglebone:~# pip install Adafruit_BBIO
```

To test the installation, launch the Python interpreter in interactive mode:

```
root@beaglebone:~# python
```

In this mode, you can key in Python code, and it will be evaluated when you hit enter. At the >>> prompt, type in the command below to attempt to import the library (keeping in mind that it's case sensitive). If you don't get any errors, you'll know you're ready to go.

```
Python 2.7.3 (default, May 29 2013, 21:25:00)
[GCC 4.7.3 20130205 (prerelease)] on linux2
Type "help", "copyright", "credits" or "license" for more information.
>>> import Adafruit_BBIO
>>>
```

You can now quit the Python interpreter with the quit() function:

```
>>> quit()
root@beaglebone:~#
```

Blinking an LED with Python

In Chapter 4 we used separate commands to turn the LED on and then turn it off. We didn't actually get the LED to blink on and off repeatedly. Using Python, you can create an infinite loop to turn the LED on and off until you terminate your script. Let's try that now.

Connect the LED

For this tutorial, we'll be using the same pins and set up as we did in Chapter 4. See "Connect an LED" on page 33 to connect an LED to pin 12 on header P8. (See Figure 5-2.)

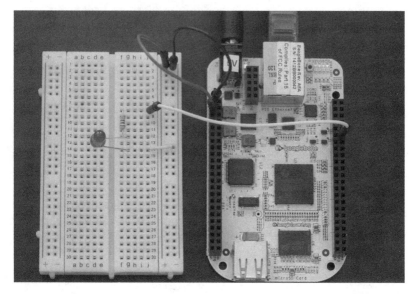

Figure 5-2. *Wiring up an LED to GPIO to pin 12 on header P8 (GPIO 44)*

Write the Code

In your home directory, use your favorite text editor to open a new file called *blink.py*. Here's how to do it in Nano:

```
root@beaglebone:~# nano blink.py
```

Getting Back to the GUI

If you prefer working with text files in a GUI on your own computer, don't worry, you're not stuck with nano. If you're connecting to your BeagleBone via SSH, you can also connect with SFTP. This will allow you to easily download files to your computer, edit them, and upload them back to the Beagle-Bone. My favorite FTP client, Transmit, even lets me edit the files in a text editor and automatically uploads them to the board whenever I hit save.

You can also plug in a keyboard, mouse, and monitor to use the BeagleBone's own GUI to edit and execute files. (See Chapter 8.)

Use the code from Example 5-1 for the contents of your file.

Example 5-1. Source code for blink.py

```
#!/usr/bin/python ❶

import Adafruit_BBIO.GPIO as GPIO ❷
import time ❸

GPIO.setup("P8_12", GPIO.OUT) ❹

while True: ❺
    GPIO.output("P8_12", GPIO.HIGH) ❻
    time.sleep(1) ❼
    GPIO.output("P8_12", GPIO.LOW) ❽
    time.sleep(1) ❾
```

❶ Tell the system to use Python when executing this file directly

❷ Allows you to use functions from Adafruit's GPIO library and refer to them with the prefix GPIO

❸ Allows you to use the Python time functions, such as sleep

❹ Set up pin 12 on header P8 to be used as an output

❺ Start an infinite loop of the code indented beneath this line

❻ Set pin 12 on header P8 as high

❼ Wait for one second

❽ Set pin 12 on header P8 as low

❾ Wait for one second

Indentation Matters!

Unlike many other languages, the indentation you use for each line of code is important to get right. It's how the Python interpreter knows what's included in part of the while loop above, for instance. You can uses spaces or tabs, just as long as you keep it consistent throughout your code.

After you've saved the file and are back at the command line, use Python to run the script:

```
root@beaglebone:~# python blink.py
```

If you see a blinking LED, you've got it right. If not, check the errors that are displayed and check your code against the code above. This script contains an infinite loop and won't end until you terminate it by hitting Control-C.

Python evaluates the code one line at a time, starting at the top of the script. When it gets to the line that says while True:, it knows to execute the indented code beneath that over and over again. The while statement can be used to test for conditions as well. For instance, if you create a game where the character has multiple lives, you may put the gameplay code underneath a line that says while lives_left > 0:. Any time the condition evaluates as true, it runs the code beneath it again, which is why while True: will run until you stop it. We're essentially saying to the Python interpreter, "blink this LED as long as true is true."

Executable Scripts

You may have noticed the first line of *blink.py* says:

```
#!/usr/bin/python
```

It might look familiar to you. It's similar to the first line of the bash script we created for the networked outlet timer in Chapter 4. This line tells the system the location of the Python interpreter to process your code. Therefore, if you'd like to execute this code without typing "python" before it, you can set *blink.py* as executable. From the command line, use the following command to do that:

```
root@beaglebone:~# chmod +x blink.py
```

Now, as long as you're in the same directory as the file, you can execute your code simply by typing:

```
root@beaglebone:~# ./blink.py
```

The "./" that proceeds the script name in the command above indicates that you're explicitly referring to the executable file in the current working directory. When executing files, you need to be explicit about their location unless you move the script to a directory that is included in your *path*. Read on for more on that.

If you'd like to be able to execute *blink.py* no matter what directory you're in, you can move it to a location that's stored in your *path*. If an executable file is inside a directory that's in your path, it is accessible wherever you are in the filesystem (you just type the filename). To view directories that are in your path, run the following command (sample output is shown):

```
root@beaglebone:~# echo $PATH
/usr/local/bin:/usr/bin:/bin:/usr/local/sbin:/usr/sbin:/sbin
```

The first directory listed in the paths above, */usr/local/bin*, is a great spot to store your script because it's where local users can store programs that aren't part of the standard distribution (this means that files you put there aren't at risk of being overwritten when you update or install packages). However, the actual directory may not already exist. That's easily rectified by creating the directory:

```
root@beaglebone:~# mkdir /usr/local
root@beaglebone:~# mkdir /usr/local/bin
```

Or, make both directories in one step by using mkdirs '-p option:

```
root@beaglebone:~# mkdir -p /usr/local/bin
```

You can then move your script to that folder:

```
root@beaglebone:~# mv blink.py /usr/local/bin
```

Now try changing to a different directory and executing *blink.py*:

```
root@beaglebone:~# cd /home
root@beaglebone:/home# blink.py
```

If everything worked, the script should execute!

Reading a Button with Python

In Chapter 4, you checked the state of a button by reading the *value* file, but didn't go much further than that. With Python, it's easy to read the state of an input and execute code based on a particular condition.

Connect the Button

For this tutorial, we'll be using the same pins and set up as we did in Chapter 4. See "Input" on page 38 to connect a button to pin 11 on header P8.

Write the Code

In your home directory, create a new file the same way you did in "Blinking an LED with Python" on page 49, but call the file *button.py*. Put the code from Example 5-2 in it.

Example 5-2. Source code for button.py

```
#!/usr/bin/python

import Adafruit_BBIO.GPIO as GPIO
import time

GPIO.setup("P8_11", GPIO.IN)   ❶

while True:
    if GPIO.input("P8_11"):   ❷
```

```
    print "The button was pressed!" ❸
  time.sleep(.01) ❹
```

❶ Set up pin 11 on header P8 to be used as an input.

❷ Read the value of pin 11; if it's high, execute the indented code beneath it.

❸ Print a message to the terminal.

❹ Avoid burdening the BeagleBone's processor by pausing briefly each time you read pin (reading it repeatedly without a pause would max out the CPU).

Try it out before we dig into the details of the code:

```
root@beaglebone:~# python button.py
```

Now when you press the button, you should see "The button was pressed!" printed on screen, probably quite a few times. Why is that? The code that's indented after while True: will execute over and over again really quickly. Each time it executes, it's checking if the button is pressed. If the button is pressed, it's printing the line to the terminal.

In fact, this block of code would be executing *even faster* if it weren't for line time.sleep(.01). Without that line, the BeagleBone's processor would be burdened with checking the state of the pin every possible moment that the processor would allow, pushing it to the absolute maximum usage. Since we're using a platform in which our code shares the processor with other programs, it's a good idea to be mindful of how hard we push it. We don't actually need to check the state of the button as fast as the processor will allow and we can still check the button fast enough to ensure that our application is responsive.

If you don't want a single button press to register multiple times, there are a few ways around this. For one, you can hold up the Python script until the button is released, like in the code in Example 5-3.

Example 5-3. Holding the process until the button is released

```
#!/usr/bin/python

import Adafruit_BBIO.GPIO as GPIO
import time

GPIO.setup("P8_11", GPIO.IN)

while True:
        if GPIO.input("P8_11"):
                print "The button was pressed!"
```

```
    while GPIO.input("P8_11"): ❶
        time.sleep(.01) ❷
    print "The button was released!"
time.sleep(.01)
```

❶   As long as the button is still pressed...

❷   ...stay in this loop, rechecking every .01 seconds.

There's also a way to look for *changes* in the value of the digital input. You can have your code look for a pin going high (rising), going low (falling), or just changing from one state to the other. These are called *interrupts*. In Example 5-4, you'll find code that has a similar effect to the code in Example 5-3, but uses interrupts instead of polling the state of the button.

Example 5-4. Using interrupts to indicate when the button is pushed or released

```
#!/usr/bin/python

import Adafruit_BBIO.GPIO as GPIO
import time

GPIO.setup("P8_11", GPIO.IN)

while True:
    GPIO.wait_for_edge("P8_11", GPIO.RISING) ❶
    print "The button was pressed!"
    GPIO.wait_for_edge("P8_11", GPIO.FALLING) ❷
    print "The button was released!"
```

❶   Wait here until the button changes from low to high.

❷   Wait here until the button changes from high to low.

Reading an Analog Input

Up until now, we've only been working with digital inputs and outputs. That is to say, only things that are on or off. For instance, the button was pressed or it wasn't, never anything in between. The same was true with the LED, we'd always write "high" or "low," never anything in between. Sometimes, however, you want an input that delivers a range of values, such as the temperature, the light levels, or the state of a dial. The BeagleBone has seven pins that can be used as analog inputs to get values from these types of sensors.

Like all computers, the BeagleBone does its work in the digital realm, so values from analog sensors must be converted to digital. The part of the BeagleBone that's responsible for this is called the *analog-to-digital converter*, or *ADC*. The ADC on the BeagleBone will let you check for values between 0 volts and 1.8 volts. (See Figure 5-3 for an analogy of analog and digital input.)

Figure 5-3. *The switch on the left is like digital: it can only be on or off. The dimmer on the right is like analog: it can be on, off, or somewhere in between.*

To try out using one of the analog input pins, you'll use a *potentiometer* or *pot*, which will act like a variable voltage divider. Essentially, it will let us turn a dial to control the amount of voltage going into the analog input pin. Here's what you'll need to try this out:

- Solderless breadboard
- Jumper wires
- 2K potentiometer

 Throughout the next few steps, be very careful that you're using the dedicated voltage and ground pins for the analog-to-digital converter (see Figure 5-5). Since it can only take up to 1.8 volts, using the other voltage pins will permanently damage your board.

Connecting a Potentiometer

1. Starting with a blank breadboard, use a jumper wire to connect pin 34 from header P9 to the ground rail on your breadboard. This is the ADC's dedicated ground pin.

2. With another jumper wire, connect pin 32 from header P9 to the positive rail in your breadboard. This is the 1.8 volt power source.

3. Insert the 2K potentiometer into the breadboard as shown in Figure 5-4.

Figure 5-4. *Connecting a potentiometer to pin 32 on header P9*

4. Out of the three pins of the potentiometer, connect the middle one to pin 33 on header P9. This is one of the seven analog input pins.

5. Connect one of the outside pins to the positive rail (choose either one for now).

6. Connect the remaining pin to the ground rail.

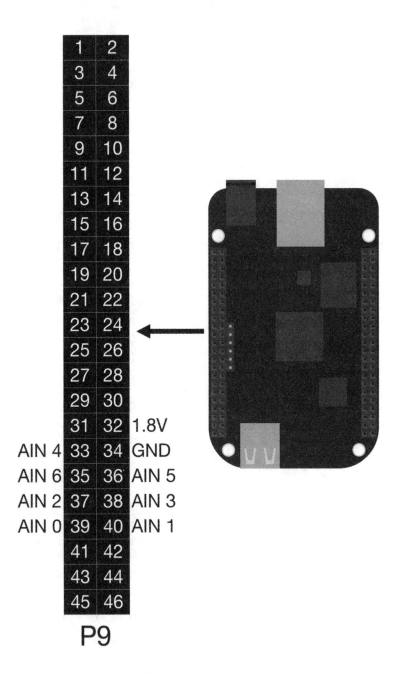

Figure 5-5. *Analog-to-digital (ADC) converter pins (BeagleBone illustration courtesy of the Adafruit Fritzing library)*

Writing the Code

Create a new file called *potentiometer.py* and put in the code from Example 5-5.

Example 5-5. Source code for potentiometer.py

```
#!/usr/bin/python
import Adafruit_BBIO.ADC as ADC ❶
import time

ADC.setup() ❷

while True:
    print ADC.read("P9_33") ❸
    time.sleep(.5)
```

❶ Allows you to use functions from Adafruit's ADC library and refer to them with the prefix ADC.

❷ Sets the pins to be used as analog inputs.

❸ Prints out a reading from the analog input pin 33 on header P9.

Execute the script and watch the terminal as you turn the pot from one side to the other:

```
root@beaglebone:~# python potentiometer.py
0.00333333341405
0.00388888898306
0.753888905048
0.997777760029
0.997777760029
0.997777760029
0.996666669846
0.484444439411
0.348333328962
0.101666666567
0.00277777784504
0.00388888898306
0.00277777784504
0.00222222227603
0.00277777784504
```

The function `ADC.read("P9_33")` returns the voltage value from pin 33 on a scale of 0 to 1. So if you want to get the actual voltage, you can multiply that number by 1.8. In Example 5-6, the reading from pin 33 is multiplied by 1.8 and then stored in a variable called `value`. That number stored in value is then rounded to two decimal places to make it look better when it's printed to the console.

Example 5-6. Source code for potentiometer.py, modified to output the voltage

```
#!/usr/bin/python
import Adafruit_BBIO.ADC as ADC
import time

ADC.setup()

while True:
    value = ADC.read("P9_33") * 1.8 ❶
    print round(value, 2) ❷
    time.sleep(.5)
```

❶   Read the value of pin 33, multiply it by 1.8 and store in memory as the variable **value**

❷   Round **value** to two decimal places and then print it to the terminal

Now run the code in Example 5-6 to output the voltage values instead of the scaled values.

```
root@beaglebone:~# python potentiometer.py
1.8
1.8
1.8
1.8
1.76
1.44
1.35
1.35
1.35
1.12
0.88
0.88
0.87
0.45
0.4
0.4
0.0
0.0
```

There are many analog sensors out there that measure things like proximity, light, temperature, acceleration, orientation, and much more. You can also find touch surfaces or arcade-style control sticks that act as analog sensors. Having analog input built into the BeagleBone is one of the reasons why it's well-suited for so many different creative hardware projects.

Analog Output (PWM)

When you first tried digital output on the BeagleBone, you turned an LED on and off. If you want to fade that LED so that it's somewhere between totally on and totally off, you can use a method called *pulse-width modulation* or PWM. There are 8 PWM channels that can be used on the BeagleBone and each channel can be connected to a few different pins as seen in Table 5-1.

Table 5-1. *PWM Channels and their respective pins. Pins marked with an asterisk are used for HDMI output on the BeagleBone Black.*

PWM Channel	Available on physical pins
EHRPWM 0A	P9_22, P9_31
EHRPWM 0B	P9_21, P9_29
EHRPWM 1A	P8_36*, P9_14
EHRPWM 1B	P8_34*, P9_16
EHRPWM 2A	P8_19, P8_45*
EHRPWM 2B	P8_13, P8_46*
ECAPPWM2	P9_28
ECAPPWM0	P9_42

 You don't need to know the PWM channel names since the Adafruit's BeagleBone IO Python Library lets you change the PWM on each pin by referring to its physical pin number. However, it's important to know that since a few different pins share the same PWM channel, changing one will change the other pins that share that channel.

Pulse-width modulation is a way of pulsing a pin on and off really quickly so that it seems like there's less voltage coming through. For instance, in the period of 1/2000th of a second, if the pin is high half the time and low the other half of the time, it will seem like there's half the voltage coming through the pin. The amount of time the pin is high is called the *duty cycle*. It can be set anywhere from 0 (totally off) to 100 (totally on).

If you use PWM with an LED at less than 100% duty cycle, it will start to appear dimmer and dimmer until you get close to 0 when it will turn off completely. PWM isn't only used for LEDs; it can also be used to control things like DC motors, hobby servos, or even make some basic tones with a speaker. In the following example, we'll keep things simple and fade an LED.

Connect the LED

Connect an LED to the BeagleBone much like you did in "Connect an LED" on page 33. However, connect the LED to pin 13 on header P8, which is one of the GPIO pins that is capable of PWM.

Write the Code

In your home directory, create a new file the same way you did in "Blinking an LED with Python" on page 49, but call the file *pwm.py*. Put the code from Example 5-7 in it.

Example 5-7. Source code for pwm.py

```
#!/usr/bin/python

import Adafruit_BBIO.PWM as PWM ❶
import time

PWM.start("P8_13", 0) ❷

for i in range(0, 100): ❸
    PWM.set_duty_cycle("P8_13", float(i)) ❹
    time.sleep(.1)

PWM.stop("P8_13") ❺
PWM.cleanup()❻
```

❶   Allows you to use functions from Adafruit's PWM library and refer to them with the prefix PWM.

❷   Start using PWM on pin 13 of header P8 and set its duty cycle to 0.

❸   Loop the code indented beneath, counting up from 0 to 100 and storing that value in i.

❹   Set the duty cycle of pin 13 on header P8 to be value of i, which increments with each iteration of this loop.

❺   Stops pulsing pin 13 of header P8.

❻   Disable PWM on any remaining pins.

Save the file and execute it from the command line:

```
root@beaglebone:~# python pwm.py
```

Now watch the LED. You should see it fade up from totally off to totally on! If you want to take this further, try modifying the code so that it fades up and down like a breathing LED.

Taking it Further

This chapter covered the basics of programming the inputs and outputs with Python. Of course, a good project usually goes beyond a simple button and LED. Hopefully, you can apply the basics in this chapter to all kinds of creative projects!

If you want to learn more about the ins and outs of Python, I recommend *Learn Python The Hard Way (http://learnpythonthehardway.org/book/)*, a free online book that walks you through all the Python fundamentals.

6/Putting Python Projects Online

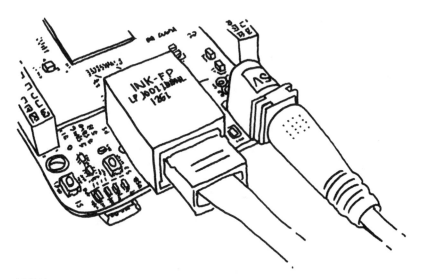

With a connection to the Internet and the slew of GPIO pins available to you, the BeagleBone makes a fantastic platform for building *Internet of Things* projects. That is, physical projects that interact with the Internet in some way. It could be as basic as sending email alerts when a physical thing happens, sending sensor data to the cloud, or even controlling a robot via a web-based interface.

In the project at the end of Chapter 2, I mention that you can change your router settings so that you can use SSH to switch the light or change its timing settings from wherever you are in the world. While using SSH to log into your BeagleBone may raise your geek cred, it's not an elegant solution for Internet-connected projects. What if your less tech-savvy house mates want to make changes to the settings? Hopefully you don't expect them to fire up a terminal and edit the BeagleBone's crontab file.

In this chapter, I'll show you a few ways to make an elegant interface to Internet-connected projects. You'll start off by setting up email notifications for physical events. Then you'll get into setting up a web server with Python so that you can have a web-based interface for reading the status of the pins. Finally, you'll set up a simple data log that can be accessed from the Web.

Sending an Email Alert

Let's say the kids in your family have a tendency to leave the back door open and you'd like to get an email alert any time the door has been left open for more than a minute. Using Adafruit's BeagleBone IO Python Library ("Installing Adafruit's BeagleBone IO Python Library" on page 48) along with Python's built-in email libraries, you can achieve this without a lot of code. Before I jump into how to connect these two resources together with code, I want to explain a Python fundamental, the *function*.

A function is a way of encapsulating a block of code meant to accomplish a single task. For instance, there are a few lines of code involved in reading a temperature sensor and printing its value in degrees Fahrenheit. You can write all that code in a function called printTemperatureInF() so that you can execute that code in just one command. You can also create a function that takes an input value in Celsius and outputs the converted value in Fahrenheit. Inputs to a function are called *parameters* and outputs are called *return values*.

Since there will be a few lines of code needed in order to connect to our mail server and send the email, we may want to encapsulate that code into its own function. If our project expands to alert us to other events, such as if the dog's water bowl is low, we don't have to have two different sets of code in order to take care of connecting to the mail server and sending the message. In other words, it's like we're saying to Python, "Here's how to send me an email." And then when we need to, we can say to Python, "OK, send me an email with the subject line 'Alert!\' and the body 'The back door is still open!\'"

Functions in Python

You may not have known it, but you were already using functions in Chapter 5. They were all part of Python's basic set of functions or in Adafruit's BeagleBone IO Python Library. For instance, the line of code GPIO.set up("P8_12", GPIO.OUT) calls a function written by Adafruit to make it easy to set pin 12 on header P8 as an output pin. There are a few steps involved in setting the mode of a GPIO pin and therefore a few lines of code. Adafruit's library makes it easy for us to do it in just one line of code. Writing your own function is actually pretty easy as seen in Example 6-1.

Example 6-1. Source code for function.py

```
def greet(name): ❶
        print "Hello, " + name + "! Welcome to my Python script!" ❷

greet("Andrew") ❸
```

❶ Define a function called greet that takes an input called name.

❷ Print a greeting message to the console with the value of name.

❸ Call the function with the name "Andrew".

When you put the code from Example 6-1 into a file and execute it, you should see the greeting printed. Try calling greeting() multiple times with different names in the same script. You'll see that you can to reuse the same code with different names quite easily.

All function definitions start with the keyword def and then the name of the function that you want to make. Inside the parentheses, you can name each of the *parameters*, or inputs, that you want the function to take. In the example above, the function greet takes a single parameter, name. Within the function, we can use that parameter's value by referencing it like we do in the print statement.

The Email Function

Let's get started writing the function that takes care of sending an email. We can test this function by itself, before getting into working on the physical side of the project. It's a good idea to compartmentalize the different parts of your project and be able to test them individually before connecting them all together.

 Be careful to not send dozens of email messages in quick succession, or you may find that your email provider slows down your access to email, or worse, suspends it temporarily. You may want to set up a separate email account for testing so you don't disrupt access to your personal email. If you need to send hundreds or thousands of emails, check out SendGrid (*http://sendgrid.com*), which offers volume email services starting at $10/month.

You'll need to find the details of your mail provider's SMTP server, which is what you connect to in order to send email. Since Gmail is a popular email service, I included their server name in Example 6-2. The code below is based on the examples found on Python's site (*http://docs.python.org/2/library/*

email-examples.html). Create a file called *emailer.py* and put the code from Example 6-2 in it (see Figure 6-1 for a visual representation).

Example 6-2. Source code for emailer.py

```
import smtplib ❶
from email.mime.text import MIMEText ❷

def alertMe(subject, body): ❸
        myAddress = "myAddress@gmail.com" ❹
        msg = MIMEText(body) ❺
        msg['Subject'] = subject ❻
        msg['From'] = myAddress ❼
        msg['Reply-to'] = myAddress ❽
        msg['To'] = myAddress ❾

        server = smtplib.SMTP('smtp.gmail.com',587) ❿
        server.starttls() ⓫
        server.login(myAddress,'my_password') ⓬
        server.sendmail(myAddress,myAddress,msg.as_string()) ⓭
        server.quit() ⓮

alertMe("Alert!", "Garage door is open!") ⓯
```

❶   Import the functions needed to connect to the SMTP mail server.

❷   Import the functions needed to create an email.

❸   Create a function called `alertMe` that takes two inputs: the email's subject and body.

❹   Create an object called `myAddress` and store your address as a string (replace `myAddress@gmail.com` with your own address).

❺   Create an object called `msg` and store the body in it in the MIME format (the body is passed as an input to this function).

❻   Set the subject of the message from the input of the function.

❼   Set the from address of the email to what you set in <4> .

❽   Set the reply-to address of the email to what you set in <4> .

❾   Set the to address of the email to what you set in <4>.

❿   Connect to SMTP server smtp.gmail.com on port 587 (replace these with your server details if not Gmail).

⓫   Start encrypting all commands to the SMTP server (you may need to remove this line for some email providers).

⓬   Send login credentials to the server (replace these with your login and password).

⓭   Send the email.

⓮   Close the connection to the SMTP server.

⑮ Call the function `alertMe` with the subject line "Alert!" and the body "Garage door is open!"

Figure 6-1. *The alertMe function in Example 6-2 takes a subject and a body as inputs and connects to the email server to send the message.*

Hardcoding to Make Things Easy

In Example 6-2 we *hardcoded* the SMTP server, login, email address, and password into the `alertMe` function, which means we put these details directly into the code as opposed to taking them as inputs from other sources. If you were working on a large-scale project with other developers and a large userbase, it wouldn't normally be good practice to put these details into the code, but rather take them as inputs to the function or pull the information from a datastore. However, as long as you're making a personal project, hardcoding this information can keep things simple.

If you execute *emailer.py* (you can run it with the command `python emai ler.py`) and don't see any errors, check your email for the alert. If you have it, you know you have a working function that you can execute within your code to send you any kind of email alert you'd like. If you're having trouble, double check your provider's email settings.

After you've got it working, remove the test line `alertMe("Alert!", "Garage door is open!")` and save the file. Now this function definition can be imported into other Python files as long as *emailer.py* is in the same directory.

The Door Sensor

Now that you have the email part of the project working, it's time to work on the physical side. Detecting whether or not a door is open is pretty simple. A door security sensor will act like a digital switch and can connect directly to the BeagleBone. These sensors can be found in a hardware store or in your local Radio Shack.

The sensors typically come in two varieties: normally open (denoted N/O) or normally closed (denoted N/C). A normally open sensor will disconnect the two terminals when the door is opened and the sensor is separated from the magnet. Conversely, a normally closed sensor will connect the two terminals when the door is opened. Some door sensors will work both ways and will have a common terminal, a normally open terminal, and a normally closed terminal, like the one in Figure 6-2. Any version will work in this project because we can use code to check for either a high or low signal.

Figure 6-2. *This door sensor has a normally open and a normally closed terminal.*

For the code below, I'll be using a normally closed sensor so that an open door will send a high signal and closed door will send a low signal. In the code, I will indicate where the logic should be changed in case you have a normally closed sensor.

1. Connect one terminal of the door sensor to pin 11 on header P8. (See Figure 6-3.)

2. Connect the other terminal of the door sensor to one of the 3.3 volt pins.

3. Connect a 10K pull-down resistor between pin 11 on header P8 and ground. (See "Input" on page 38 for a refresher on pull-down resistors.)

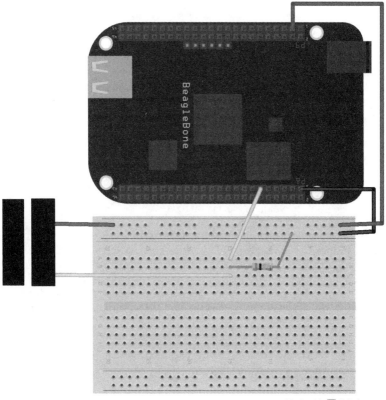

Made with 🅵 Fritzing.org

Figure 6-3. *Wiring the door sensor to pin 11 on header P8 with a 10K pull-down resistor.*

The Code

It can be a bit confusing when you start working with lots of nested loops and if statements, so let's first map out the logic in *pseudocode*, a way of using plain language to describe what code is going to do.

- Set the value of alertTime to 0.
- Loop forever:
  - Check if the door is open.
  - If the door is open:
    - Check if alertTime is 0.
    - If alertTime is 0:
      - Set the value of alertTime to the current time plus 60 seconds.
    - If alertTime is not 0:
      - Check if the current time is past alertTime.
      - If the current time is past alertTime:
        - Send the alert email.
        - Wait here until the door is closed.
      - If the current time is not past alertTime:
        - Do nothing.
  - If the door is not open:
    - Set the value of alertTime to 0.

The main loop constantly checks if the door has been opened. When it senses that the door has gone from being closed to being opened, it saves a time value into alertTime. That time is 60 seconds in the future. For every subsequent check, if the door is still opened, it checks if 60 seconds has passed. If so, it sends an email alert. If the door has been closed, it clears the value of alertTime, ensuring that the next time the door is open, it will start tracking a full 60 seconds again.

There's also a loop after the email alert is sent to ensure that only a single email is sent if the door has been open for more than a minute. Example 6-3 shows how that logic is converted into Python.

Example 6-3. Source code for doorAlert.py

```
import time
import Adafruit_BBIO.GPIO as GPIO
from emailer import alertMe ❶

GPIO.setup("P8_11", GPIO.IN)

alertTime = 0 ❷

while True:
    if GPIO.input("P8_11"): ❸
        if alertTime == 0: ❹
            alertTime = time.time() + 60 ❺
```

```
        print "Door opened. Alerting in 60 seconds if left open."
    else: ❻
        if time.time() > alertTime: ❼
            alertMe("Alert!", "The door has been left open!") ❽
            print "Door has been open for 60 seconds. Sent email alert!"
            while GPIO.input("P8_11"): ❾
                time.sleep(.1)
    else:
        alertTime = 0 ❿
    time.sleep(.01)
```

❶   Import the alertMe function from *emailer.py*.

❷   Set the value of alertTime to 0 initially.

❸   If the door is opened, execute the indented code below (If using a normally open sensor, this line should read if not GPIO.in put("P8_11"):).

❹   If the time to send the alert has not yet been set as the value of alertTime...

❺   ...then take the time now, add 60 seconds and store that in alertTime.

❻   If the time to send the alert has been set, execute the code indented below.

❼   If the current time is past alertTime...

❽   Call the alertMe function from *emailer.py* with the subject and body of the email.

❾   Wait here until the door is closed again to prevent multiple emails being sent (If using a normally open sensor, this line should read while not GPIO.input("P8_11"):).

❿   If the door is closed, set alertTime back to 0.

The *doorAlert.py* file must be in the same directory as *emailer.py* in order to import the alertMe function from that file.

You must remove the line from *emailer.py* that sends the test email otherwise you will receive an email each time you start *doorAlert.py*.

Next you'll want to set up *doorAlert.py* to execute in the background after your BeagleBone starts up. If you want to do that now, jump ahead to Appendix B.

Web Interface

Back in the early days of the Web, every page you visited was represented by a HTML file on the web server. When you sent a request for a particular URL, the server found that file in its directory structure and transmitted it to you. While many web servers still operate in that fashion, there's a new generation of web servers out there. When you request a URL, they dynamically generate the HTML and transmit it to you based on rules found within the code.

One of the things I absolutely love about embedded Linux platforms like the BeagleBone is that they can run many modern web application frameworks. You can use these frameworks along with the GPIO libraries to make web interfaces for physical projects. For instance, let's say that instead of an email alert when the door is left open, you just want to be able to view a web page to see if the door is open or closed.

First Steps with Flask

For this project, you'll use the Flask framework (*http://flask.pocoo.org/*) for Python to serve the web page. And since you're still working in Python, you'll continue to use Adafruit's BeagleBone IO Python Library. Before you get into creating your own dynamic site, you'll have to install Flask and its dependencies using `pip`:

```
root@beaglebone:~# pip install flask
```

To test out Flask, you'll create a simple web server that replies "Hello, world!" to the request it receives. Start by creating a new file called *hello-flask.py* with the code from Example 6-4.

Example 6-4. Source code for hello-flask.py

```
from flask import Flask ❶
app = Flask(__name__) ❷

@app.route("/") ❸
def hello():
    return "Hello, world!" ❹

if __name__ == "__main__": ❺
    app.run(host='0.0.0.0', port=81, debug=True) ❻
```

❶   Import the Flask library's functions.

❷   Create a flask object called "app".

❸   When Flask gets a request for / (the web root), run the function beneath.

❹   Send the text "Hello, world!" to the web browser.

If this file itself is being executed (and not imported by another file), execute the code indented beneath it.

 Have the flask server run on port 81 and report and log requests and errors to the console.

After saving *hello-flask.py* and executing it, you should see the following on the terminal:

```
root@beaglebone:~# python hello-flask.py
 * Running on http://0.0.0.0:81/
 * Restarting with reloader
```

Then try opening your web browser and navigating to `http://beagle bone.local:81`. If everything worked, you should see "Hello, world!" in your browser window. If you look back at your BeagleBone terminal, you'll see the following:

```
10.0.1.5 - - [16/Jul/2013 12:32:20] "GET / HTTP/1.1" 200 -
10.0.1.5 - - [16/Jul/2013 12:32:20] "GET /favicon.ico HTTP/1.1" 404 -
```

The first line represents the GET request from the browser for /. The server responded with a code 200, which means OK. This response appeared when the text "Hello, world!" was transmitted to the browser. The second line is another GET request for */favicon.ico*, to which the server responded with code 404, meaning "file not found". Browsers request a small graphic icon when visiting a site to represent it in the location bar and in the bookmarks menu. Your site does not have a *favicon.ico* file, so you can ignore this error. It won't cause any problems.

To terminate the server and get back to the command line, type Control-C.

Templates with Flask

Example 6-4 sent only text to the web browser, none of the HTML formatting that you'd normally expect from a web page. You could write the code yourself to send this formatting along with the `return` statement in that example, but including all that code in the Python script would get unwieldy very quickly. Luckily for us, Flask uses a *template engine* called Jinja2 (*http:// jinja.pocoo.org/*), which takes care of combining HTML formatting with the data that our application will provide.

With the capabilities of Jinja2, you can create HTML files that have place-holders for data that comes from our web application. This means that the code that controls the look of the page is kept separate from the actual logic of the application. Flask's `render_template` function will take in the data and HTML template filename, pass it to Jinja2, which will return the properly for-matted HTML to Flask to return to the user.

To try it out yourself, create a new file called *flask-template.py* and place the code from Example 6-5 in it. Then create a subdirectory called *templates* and create a file in there called *main.html*. Use Example 6-6 as the contents for that file.

Example 6-5. Source code for flask-template.py

```
from flask import Flask, render_template ❶
app = Flask(__name__)
import datetime

@app.route("/")
def hello():
        now = datetime.datetime.now()  ❷
        timeString = now.strftime("%Y-%m-%d %H:%M") ❸
        templateData = {
                'title': 'Hello!',
                'time': timeString
        } ❹
        return render_template('main.html', **templateData) ❺

if __name__ == "__main__":
        app.run(host='0.0.0.0', port=81, debug=True)
```

❶ Import the `render_template` function.

❷ Store the current time into `now`.

❸ Create a human-readable string of the date and time.

❹ Create a dictionary of values to pass to the template (in key:value format).

❺ Render the template HTML from `templates/main.html` using the values from the `templateData` dictionary.

Example 6-6. Source code for templates/main.html

```
<!DOCTYPE html>
<head>
<title>{{title}}</title> ❶
</head>

<body>
<h1>Hello, world!</h1>
<h2>The date and time on the server is: {{ time }}.</h2> ❷
</body>
</html>
```

❶ Take the value of the key `title` and replace it here.

❷ Take the value of the key `time` and replace it here.

In Example 6-5, the line that says `return render_template('main.html', **templateData)` instructs Flask to use the Jinja2 template engine to prepare the page. It will look in the folder *templates* and find the specified file, in this case, *main.html*. Flask also gives Jinja2 the values stored in the dictionary *templateData*, which was created a few lines above. These values can be placed in the template by name in between two curly braces, as you see in Example 6-6.

After saving *flask-template.py* and *templates/main.html*, execute the script from the command line:

```
root@beaglebone:~# python flask-template.py
 * Running on http://0.0.0.0:81/
 * Restarting with reloader
```

Now when you navigate to to `http://beaglebone.local:81`, you should see the BeagleBone's time in the browser.

Combining Flask and GPIO

Using the same circuit that you did in "The Door Sensor" on page 70, you can create a web interface with Flask to check the status of the door (see Examples 6-7 and 6-8). Every time a request is made from the server, it will check the state of the GPIO pin and pass "open" or "closed" strings into the template, which is then sent to the browser making the request.

Example 6-7. Source code for flask-door.py

```
from flask import Flask, render_template
app = Flask(__name__)
import Adafruit_BBIO.GPIO as GPIO  ❶

GPIO.setup("P8_11", GPIO.IN)

@app.route("/")
def hello():
        if GPIO.input("P8_11"):
                doorStatus = "open"  ❷
        else:
                doorStatus = "closed"  ❸
        templateData = {
                'doorStatus': doorStatus,
        }  ❹
        return render_template('main-door.html', **templateData)  ❺
if __name__ == "__main__":
        app.run(host='0.0.0.0', port=81, debug=True)
```

❶   Import the GPIO functions from Adafruit's library.

❷   If the door is open, set the **doorStatus** string to "open".

❸ Otherwise, set the doorStatus string to "closed".

❹ Store the doorStatus string in a dictionary with the key doorStatus.

❺ Be sure that this references the correct template HTML file, in this case main-door.html.

Example 6-8. Source code for templates/main-door.html

```
<!DOCTYPE html>
<head>
<title>Matt's Apartment</title>
</head>

<body>
<h1>Matt's Apartment</h1>
<h2>The door is {{ doorStatus }}.</h2> ❶
</body>
</html>
```

❶ When the HTML is generated by the BeagleBone, it will substitute {{ doorStatus }} with the string determined in flask-door.py

Going Further with Flask

Following the code from Example 6-7, you can use the BeagleBone to host web services for additional digital inputs. You can also use Flask to control digital outputs, pulse the PWM pins, and get readings from analog sensors connected to the analog-to-digital converter.

Another powerful feature of Flask is that it can process variables within URLs. For instance, you can set the brightness of an LED based on what URL you request from your browser. You can set up Flask so that the address *http://beaglebone.local:81/ledLevel/100* would set the LED to be fully bright and *http://beaglebone.local:81/ledLevel/50* would set it to be half dim. All this can be done with just a little bit of code as shown in Example 6-9.

Example 6-9. Source code for flask-door.py modified to add PWM route

```
from flask import Flask, render_template
app = Flask(__name__)
import Adafruit_BBIO.GPIO as GPIO
import Adafruit_BBIO.PWM as PWM

PWM.start("P8_13", 0.0)

@app.route("/")
def hello():
```

```
    if GPIO.input("P8_11"):
        doorStatus = "open"
    else:
        doorStatus = "closed"
    templateData = {
        'doorStatus': doorStatus,
    }
    return render_template('main-door.html', **templateData)

@app.route('/ledLevel/<level>') ❶
def pin_state(level):
    PWM.set_duty_cycle("P8_13", float(level)) ❷
    return "LED level set to " + level + "."

if __name__ == "__main__":
    app.run(host='0.0.0.0', port=81, debug=True)
```

❶   Pass the variable `level` from the URL into the function.

❷   PWM pin 13 on header P8 according to the level in the URL.

After adding the route above to your web application, your server will respond to requests made to *http://beaglebone.local:81/ledLevel/50* and will PWM the LED at 50% duty cycle so that it's dim.

Data Logging with Xively

Since it has a traditional filesystem built right in, using the BeagleBone makes a lot of sense at the core of of data logging projects. The data it collects can be saved to a file and then accessed via SFTP, viewed via the Web, or emailed to you directly. You can also have the BeagleBone push data to a cloud service like Xively (*https://xively.com/*), which makes it easy to view current and historical data from sensors via the web (see Figure 6-4).

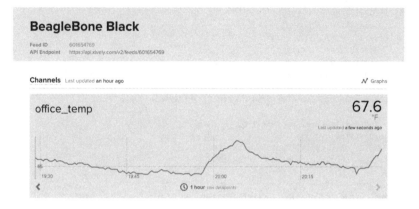

Figure 6-4. *The web view of a Xively data feed*

In this section, you'll learn how to hook up an analog temperature sensor to the BeagleBone, use Python to calculate the temperature from the voltage readings, and then upload the data to Xively.

Connecting the Temperature Sensor

In addition to an Internet-connected BeagleBone, you'll need the following to prototype this project:

- Solderless breadboard
- Jumper wires
- TMP35 analog temperature sensor (This component is available from vendors like Digi-Key or Mouser. You can also use a TMP36. See Using a TMP36 Sensor for more information.)

Since it's an analog sensor, you'll connect the TMP35 to one of the analog input pins on the BeagleBone. Pin 1 of the TMP35 connects to 3.3 volts, pin 2 connects to analog in at pin 40 on header P9 (or any of the other analog input pins) and pin 3 connects to ground, as shown in Figure 6-5. The TMP35 will output 0.1 to 1.2 volts for temperatures from 10 to 120 degrees Celsius. In fact, the formula to determine the temperature is:

degrees celsius = output voltage in mV / 10

To test the circuit, use the code in Example 6-10 in a new file called *tmp35.py*.

Example 6-10. Source code for tmp35.py

```
import Adafruit_BBIO.ADC as ADC
import time
```

```
ADC.setup()

while True:
    mV = ADC.read('P9_40') * 1800 ❶
    celsius = mV / 10 ❷
    fahrenheit = (celsius * 9/5) + 32 ❸
    print round(fahrenheit, 1) ❹
    time.sleep(1)
```

❶ Store the analog reading from pin 40 on header P9 into `reading` (converting it to millivolts).

❷ Convert from the milivolts measurement to Celsius.

❸ Convert from Celsius to Fahrenheit.

❹ Print the Fahrenheit value, rounded to a single decimal place.

When you execute this Python script, you should see a Fahrenheit value printed every second. Try warming the sensor with your fingers. The value should go up! Let go and watch the value drop again.

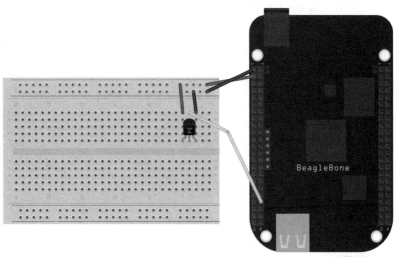

Made with **Fritzing.org**

Figure 6-5. *Wiring the TMP35 temperature sensor to the analog pin at pin 40 on header P9.*

Connecting to Xively

Xively is an *Internet of Things* platform that takes in data from sensors all over the world and stores that in a database. It can then serve the data via their web interface or via an API for your own applications. It was previously known as Cosm, and before that, Pachube. While it's possible to configure the BeagleBone to store and serve the data itself, the advantage to using Xively is that you can access the data wherever you have an Internet connection without needing to do any special network configuration.

Additionally, their developer libraries make it very easy to connect to their services in many different programming languages. In this section, you'll use their Python library to push analog sensor data to Xively.

1. Create an account at the Xively signup page (*https://xively.com/sign up/*). A free developer account will be sufficient for personal use.

2. Log in to your account and click "Develop" on the top of the page.

3. Click "Add Device"

4. Give your device a name and choose whether you want the data to be public or private. You can change these settings later, if you wish. After clicking "Add Device," you'll be taken to the device's page.

5. From the device's page, click "Add Channel" to create a new sensor channel such as "office_temp." The other fields are optional.

6. Take note of the auto-generated API key, the Feed ID, and your channel's name on the device's page. You'll need to put those in your Python code.

7. On your BeagleBone, use `pip` to install the Xively Python library:

```
root@beaglebone:~# pip install xively-python
```

8. Create a new file called *xively-temp.py* with the contents of Example 6-11. Be sure to replace the API key, feed ID, and channel name with information from your own account.

9. When you execute *xively-temp.py*, it will take the temperature reading and send it to Xively every 20 seconds. Check the device's page on Xively.com to see the data start coming in!

Example 6-11. Source code for xively-temp.py

```
import Adafruit_BBIO.ADC as ADC
import time
import datetime
import xively
from requests import HTTPError ❶

api = xively.XivelyAPIClient("API_KEY_HERE") ❷
feed = api.feeds.get(FEED_ID_HERE) ❸

ADC.setup()

while True:
    mV = ADC.read('P9_40') * 1800
```

```
celsius = mV / 10
fahrenheit = (celsius * 9/5) + 32
fahrenheit = round(fahrenheit, 1)

now = datetime.datetime.utcnow() ❹
feed.datastreams = [
    xively.Datastream(id='office_temp', current_value=fahrenheit, at=now)
❺
]

try:
    feed.update() ❻
    print "Value pushed to Xively: " + str(fahrenheit)
except HTTPError as e: ❼
        print "Error connecting to Xively: " + str(e)
time.sleep(20)
```

❶ Import **HTTPError** so that we can gracefully process possible errors when connecting to Xively to send data.

❷ Enter your API key within quotation marks here.

❸ Enter your feed ID within parenthesis here.

❹ Store the current time in the variable now.

❺ Enter your channel ID here.

❻ Push the data to the Xively server.

❼ If there's an error contacting the Xively server, print a message to the terminal, but try again in 20 seconds.

Using a TMP36 Sensor

Example 6-11 contains code to convert the analog output from the TMP35 to the temperature in degrees Celsius. It's also possible to use the popular TMP36 temperature sensor, but you'd have to adjust the conversion to `celsius = (mV - 500) / 10`. Be aware that the TMP36 outputs values greater than the ADC's limit of 1.8v at temperatures above 130 degrees Celsius (266 degrees Fahrenheit).

Under a single device within Xively, you can also add more channels to send multiple sensor values. After storing a second value from the analog-to-digital converter, add the value to the **feed.datastreams** list following this pattern:

```
feed.datastreams = [
    xively.Datastream(id='office_temp', current_value=fahr, at=now),
    xively.Datastream(id='outdoor_temp', current_value=outside_fahr, at=now)
]
```

As it stands, Example 6-11 will only run when you execute it from the command line and it will stop when you terminate it, when you log out, or when you reboot the BeagleBone. Clearly, this isn't the desired mode of operation for most projects. To set up this project so that it launches automatically, flip ahead to Appendix B.

Taking it Further

It would be difficult to cover all of the possible ways your BeagleBone could communicate over the Internet. However, since Python is such a widely used language, it's not difficult to find a way of doing things such as sending tweets, checking the weather, or pulling data from Google Maps. Here are a few additional resources for using Python online:

Requests Library (http://docs.python-requests.org/en/latest/)
This is a popular library helps you build HTTP requests and process their responses.

Temboo (https://temboo.com/)
Temboo makes it easy to connect to hundreds of different APIs.

Pusher (http://pusher.com/)
Pusher is a cloud service and API that helps you send real-time information to web pages. For instance, you can create a web page that always shows a live-updating temperature reading from the BeagleBone.

7/Bonescript

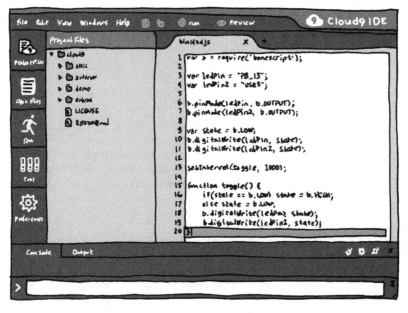

As a programming language, JavaScript has come a long way since its debut as a tool for adding client-side interactivity to websites. It's still responsible for much of the roll-over graphics, form validation, and asynchronous communication with the server (when you send data to or receive it from a website without leaving the page). Chances are, most of the sites you visit use JavaScript in some way or another.

These days, it's used for much more than just as a language for client-side (in-browser) scripting. With a framework like Node.js (*http://nodejs.org/*), JavaScript can act as the engine of a web server, which affords web application developers many of the advantages of the event-driven nature of the language. Instead of executing one line of code and waiting for it to finish before executing the next, much of Node.js's functions are *non-blocking*, meaning they can handle other things while waiting for a task to complete.

The strengths of JavaScript can also help out in the realm of hardware projects. BoneScript (*https://github.com/jadonk/bonescript*) is a Node.js library that brings a lot of the pin control functionality into JavaScript and it will look familiar to anyone with Arduino experience because it has functions like digitalRead, digitalWrite, analogRead, analogWrite, and a few others. However, these functions were designed to take advantage of the event-driven nature of JavaScript.

To explain, when you coded with Python and Adafruit's BeagleBone IO Python Library in Chapter 5, the Python interpreter executed a single line and waited for it to complete before moving on to the next line. The JavaScript engine, on the other hand, won't always wait for a line of code to complete before moving onto the next. That's why many functions have *callback* methods which tells a function to call another function when it's done its work.

The Cloud9 IDE

If you've worked with the Arduino before, you're probably familiar with the Arduino IDE, or *integrated development environment*. It's the application on your computer where you write the code, compile it, and upload it to the board. BeagleBone hosts its own web-based IDE called Cloud9 (*https://github.com/ajaxorg/cloud9/*) for writing applications in BoneScript (see Figure 7-1). To get to the IDE, you'll have to use a web browser to connect to port 3000 of your BeagleBone. So if you haven't changed the hostname from beaglebone, you'll connect to *http://beaglebone.local:3000/*. See "Changing the Hostname" on page 28 for more information. It may take a few moments after booting your BeagleBone for the Cloud9 server to respond.

Under the Project Files panel on the left side, you'll notice a directory named cloud9. Within that directory you'll see a few subdirectories. If you want to explore on your own, there are a few BoneScript examples in the demo directory. All of these files are located within the directory /var/lib/cloud9 on your BeagleBone, in case you prefer to use your own IDE or text editor to explore the examples.

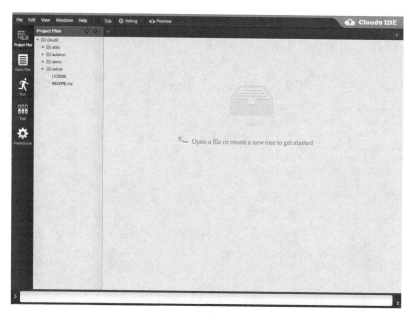

Figure 7-1. *Cloud9, a web-based IDE which is great for writing applications with BoneScript*

Beginning BoneScript

Because JavaScript is an entirely different language than Python, keep an eye on the differences in syntax. Unlike Python, JavaScript doesn't care about how you use indentation. You can even put all your code on a single line if you wanted to. (But I wouldn't recommend it!) JavaScript does however expect a semicolon after each statement.

The best way to learn the language is by example, so let's jump right in.

Blinking an LED

Double click on the file `blinkled.js` within the `demo` directory and you'll see the code from Example 7-1 appear in the editor pane. If you'd like to see an external LED blink, wire up an LED to pin 13 on header P8 (similar to what did in "Connect an LED" on page 33 but to pin 13 on P8 instead of pin 12). The code will toggle that pin and it will also toggle the on-board LED labeled USR3, which is right next to the Ethernet port (so if you don't wire up the LED, you can still see the onboard LED blink).

Example 7-1. BoneScript demo file `blinkled.js`

```
var b = require('bonescript'); ❶

var ledPin = "P8_13"; ❷
var ledPin2 = "USR3"; ❸

b.pinMode(ledPin, b.OUTPUT); ❹
b.pinMode(ledPin2, b.OUTPUT); ❺

var state = b.LOW; ❻
b.digitalWrite(ledPin, state); ❼
b.digitalWrite(ledPin2, state); ❽

setInterval(toggle, 1000); ❾

function toggle() { ❿
    if(state == b.LOW) state = b.HIGH; ⓫
    else state = b.LOW; ⓬
    b.digitalWrite(ledPin, state); ⓭
    b.digitalWrite(ledPin2, state); ⓮
}
```

❶   Load the BoneScript module, which you'll refer to as **b** throughout the code.

❷   Create a variable called `ledPin` containing the string "P8_13" to refer to the GPIO pin 13 on header P8.

❸   Create a variable called `ledPin2` containing the string "USR3" to refer to the on-board LED labeled USR3.

❹   Set `ledPin` as an output.

❺   Set `ledPin2` as an output.

❻   Create a new variable called `state` and store the value LOW in it.

❼   Write the value from `state` (LOW) to `ledPin`.

❽   Write the value from `state` (LOW) to `ledPin2`.

❾   Execute the toggle function every 1000 milliseconds (1 second).

❿   Declare a new function called `toggle`, which consists of the code between the following curly braces.

⓫   If the variable `state` contains the value LOW, set it to HIGH.

⓬   Otherwise, set the value of `state` to LOW.

⓭   Write the value of `state` to `ledPin`.

⓮   Write the value of `state` to `ledPin2`.

If the text next to the green play button on Cloud9's toolbar says Debug, pull down the button's menu and deselect "Run in debug mode" (Figure 7-2). This will change the text of the button from Debug to Run, which will ensure that the output text will be displayed at the bottom of Cloud9 every time you click Run.

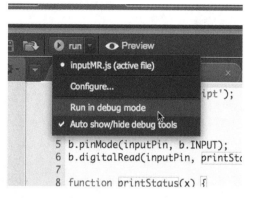

Figure 7-2. *Turning off debug mode in Cloud9*

When you click the Run button at the top of the Cloud9 editor, you'll see the LED USR3 blinking on the board and if you hooked up an LED to pin 13 on header P8, you should also see that flash in sync with the on-board LED. Press "Stop" to end the process.

What's interesting about Example 7-1 is that it creates a new function called `toggle` and then uses `setInterval` to execute it once every second. Using this method, you can set up multiple functions to run independently, but able to access the same variables and data if you wish.

Reading a Digital Input

To learn about reading a digital input with BoneScript, create a new file called `inputPrint.js` with the contents of Example 7-2. In Cloud9, it's as easy as clicking the File menu and selecting New File. When you're done inputting the code, save it (File→Save) and you will be prompted for a name (*input-print.js*) and location.

Just like you did in "Input" on page 38, wire up a button or switch to pin 11 on header P8.

Example 7-2. Source code for `inputPrint.js`

```
var b = require('bonescript');

var inputPin = "P8_11"; ❶

b.pinMode(inputPin, b.INPUT); ❷
b.digitalRead(inputPin, printStatus); ❸
```

```
function printStatus(x) { ④
    if (x.value == b.HIGH) { ⑤
        console.log("The pin is HIGH");
    }
    else { ⑥
        console.log("The pin is LOW");
    }
}
```

①   Create a variable called `inputPin` containing the string "P8_11" to refer to the GPIO pin 11 on header P8.

②   Set `inputPin` as an input.

③   Read the state of `inputPin` and pass that to the function `printStatus`.

④   Create a function called `printStatus` that takes in a parameter that will be referred to as x.

⑤   If the value of the pin is high, write "The pin is HIGH" to the console.

⑥   Otherwise, write "The pin is LOW" to the console.

When you click Run, the code will be executed once and will display the state of the button or switch in the Output tab at the bottom of Cloud9. Try changing the state of the button or switch and clicking Run again. If you're using a momentary switch, be sure to hold the button down (or keep it released) while you click Run.

Example 7-2 also demonstrates the callback functionality of the BoneScript library. The `digitalRead` function can take two parameters (or inputs):

`digitalRead(pin, callback)`

pin

    The pin number that you want to read

callback *(optional)*

    The name of the function to execute when reading the pin has completed. `digitalRead` will pass the state of the pin to this function.

The function `printStatus` created in Example 7-2 will accept a single parameter, or input, designated by the x in parentheses next to the function's name. Within the function, that input will be referred to as x. When `digitalRead` executes printStatus, it will pass `x.value` or `x.error` so that printStatus can take action based on the state of the pin (or any errors that may have occurred).

Interrupts

It's also possible to attach an interrupt to an input pin, which will allow you to monitor the pin for changes and execute a function if a particular change is detected. You can specify if you want the code to be executed when the pin is rising (it goes from low to high), falling (it goes from high to low), or both. Attaching an interrupt will allow your code to do other things without constantly needing to poll the state of the pin for changes.

The code in Example 7-3 combines code from both Example 7-1 and Example 7-2 and is meant to show that you can have the LED blink and handle changes from the input pin at the same time. This is one of the big advantages to the event-driven nature of JavaScript.

Example 7-3. Source code for `interrupt.js`

```
var b = require('bonescript');

var ledPin = "USR3";
var inputPin = "P8_11";

b.pinMode(ledPin, b.OUTPUT);
b.pinMode(inputPin, b.INPUT);

b.attachInterrupt(inputPin, true, b.CHANGE, printStatus);  ❶

var state = b.LOW;
b.digitalWrite(ledPin, state);

setInterval(toggle, 1000);

function toggle() {
    if(state == b.LOW) state = b.HIGH;
    else state = b.LOW;
    b.digitalWrite(ledPin, state);
}

function printStatus(x) {
    if (x.value == b.HIGH) {
        console.log("The pin is HIGH");
    }
    else {
        console.log("The pin is LOW");
    }
}
```

 Execute the `printStatus` function when any change is detected from `inputPin`.

When you run the code, you'll see that the USR3 led will be blinking and every time you change the state of the button or switch, a message will be printed to the console at the bottom of Cloud9.

The `attachInterrupt` function can take four parameters:

```
attachInterrupt(pin, handler, mode, callback)
```

pin
> The pin that you want to listen to.

handler
> If set as **true**, it will always execute the callback function when a state change is detected. Otherwise, you can conditionally execute the callback function.

mode
> What types of state changes to monitor for. This can be RISING (low to high), FALLING (high to low), or CHANGE (any state change).

callback *(optional)*
> The name of the function to execute when a state change is detected.

Analog Input

If you connect a potentiometer or other analog sensor to the analog input pin 32 on header P9 as instructed in "Connecting a Potentiometer" on page 57, you can also get readings from it in BoneScript. Example 7-4 shows how to take analog readings from sensors and write their values to the console.

Example 7-4. Source code for analogInput.js

```
#!/usr/bin/node

var b = require('bonescript');

var inputPin = "P9_32";

loop(); ❶

function loop() { ❷
    b.analogRead(inputPin, printValue); ❸
    setTimeout(loop, 1000); ❹
}

function printValue(x) { ❺
    console.log(x.value); ❻
}
```

❶   Execute the function **loop** for the first time.

❷   Define a function called **loop**.

❸   Take the analog reading from **inputPin** and then pass it to the function **printValue**.

❹   Execute **loop** again in 1,000 miliseconds (1 second).

❺   Define a function called **printValue** that takes an input which will be referred to as **x** within the function.

❻   Print the input's value to the terminal.

This example also demonstrates another way to set up a loop that executes repeatedly, but the effect is slightly different than in Example 7-1. The `loop` function in Example 7-4 includes the code to schedule itself to run again one second after the code before it has been executed, rather than just executing the loop once per second (no matter how long it takes to execute the code in the loop).

It would be a good idea to use the looping method in Example 7-4 if you think the code in the loop function may take longer than the interval to execute.

The `analogWrite` function can take two parameters:

`analogWrite(pin, callback)`

pin
> The pin that you want to read

callback *(optional)*
> The name of the function to execute when reading the pin has completed. `analogWrite` will pass the analog value of the pin to this function.

PWM

As discussed in "Analog Output (PWM)" on page 61, only certain pins can be used for `analogWrite` functions. Be sure to review Table 5-1 when choosing what pins to use for PWM. Example 7-5 uses pin 13 on header P8.

Example 7-5. Source code for `analogWrite.js`

```
var b = require('bonescript');

var ledPin = "P8_13";

b.pinMode(ledPin, b.OUTPUT);

b.analogWrite(ledPin, 0.05); ❶
```

❶   Set the duty cycle of `ledPin` to 5%

The `analogWrite` function can take four parameters:

`analogWrite(pin, value, frequency, callback)`

pin
> The pin that you want to use for PWM

value
> The duty cycle of the pin, between 0 (always off) and 1 (always on)

frequency *(optional)*
> The frequency of the PWM cycle in Hz (cycles per second)

callback *(optional)*
> The name of the function to execute the PWM value is successfully written

Playing with PWM: "Breathing" LED

If you want to play around with that LED a little and give it a "breathing" effect, check out Example 7-6. It uses `setInterval` to run a function repeatedly to determine how to change the LED's PWM duty cycle. That function then executes BoneScript's `analogWrite` function to set the LED's level accordingly.

Example 7-6. Source code for breathingLED.js

```
var b = require('bonescript');

var ledPin = "P8_13";
var fadingUp = true; ❶
var level = 0.0; ❷

b.pinMode(ledPin, b.OUTPUT);

b.analogWrite(ledPin, level);

changeLevel(); ❸

function changeLevel() { ❹
    if (level > 1.0) {
        fadingUp = false; ❺
    }
    if (level < 0) {
        fadingUp = true; ❻
    }

    if (fadingUp) {
        level = level + 0.01; ❼
    }
    else {
        level = level - 0.01; ❽
    }
    b.analogWrite(ledPin, level); ❾

    setTimeout(changeLevel, 10); ❿
}
```

❶ Create a variable to indicate that it'll start by fading up (when set to false, it will indicate it's fading down).

❷ Create a variable called `level` to store the PWM duty cycle and start it at 0.0.

❸ Run the `changeLevel` function (defined next).

❹ Create a new function called `changeLevel`.

❺ If `level` has exceeded 1.0, set `fadingUp` to false.

❻ If `level` has gone below 0, set `fadingUp` to true.

❼ If `fadingUp` is true, add .01 to `level`.

❽ If `fadingUp` is false, subtract .01 from `level`.

❾ Set the PWM level of `ledPin` to `level`.

❿ Execute the `changeLevel` function again in 10 milliseconds.

When you run the code, should see the LED you connected to pin 13 fading up and down! Try changing the interval for different effects.

Running JavaScript Files from the Command Line

If you'd like to execute your JavaScript code from the command line, you'll use the command **node** followed by the name of the file. For example, here's how to run the `blinkled.js` example in the BoneScript demo folder:

```
root@beaglebone:~# cd /var/lib/cloud9/demo/
root@beaglebone:/var/lib/cloud9/demo# node blinkled.js
```

To end the process, type Control-C.

Knowing how to execute the file from the command line will come in handy for setting your code up as a cron job (see "A Crash Course in Cron" on page 44) or as a system service (see Appendix B).

Setting Scripts as Executable

If you'd like to execute a script by simply typing its name, put the following on the first line of your script. This will instruct the system to look for the Node interpreter when executing the file.

```
#!/usr/bin/node
```

You can then set the file as executable:

```
root@beaglebone:~# chmod +x myScript.js
```

When you're in the directory with the file, you can now execute the file just by typing its name after **./**:

```
root@beaglebone:~# ./myScript.js
```

If you'd like to be able to execute the file no matter what directory you're in, see "Executable Scripts" on page 52.

Setting JavaScript Files to Run Automatically

You can also have your code start up automatically by adding it to the autor un directory within the cloud9 directory. A system service continually scans this directory for any JavaScript files and will run them. So as soon as you drop a file with the .js extension, it will use node to execute it and ensure that it is executed whenever you boot your BeagleBone.

Removing the file from the autorun directory will terminate the process.

BoneScript Reference

The full reference for BoneScript (see Figure 7-3) can always be accessed by going to *http://beaglebone.local/Support/BoneScript/* in your web browser (if necessary, replace beaglebone with your board's hostname). Not only does it contain reference information for each BoneScript function, but it also lets you modify and execute the code samples directly from the web page. Reference information can also be found on BeagleBoard.org's site (*http://beagleboard.org/support/bonescript/*).

Figure 7-3. *BoneScript's interactive reference pages*

8/Using the Desktop Environment

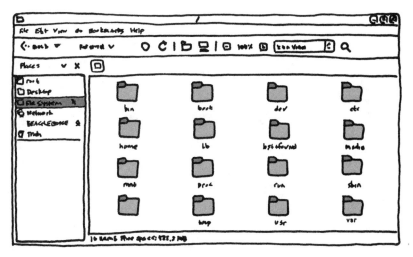

Just like a typical computer, the BeagleBone has a graphical desktop environment that you can navigate with your mouse. It will let you click through the filesystem, drag and drop files, and launch windowed graphical programs. If you're pretty good at getting around Windows or OS X, you'll have no problem working with GNOME, the desktop environment preloaded onto the BeagleBone's Ångström distribution of Linux. (See Figure 8-1.)

Much of what you have done on the command line can also be done within the desktop environment. Unless you're a command line power user, it's probably much easier to switch between files, applications, and copy and paste text when you're within the desktop.

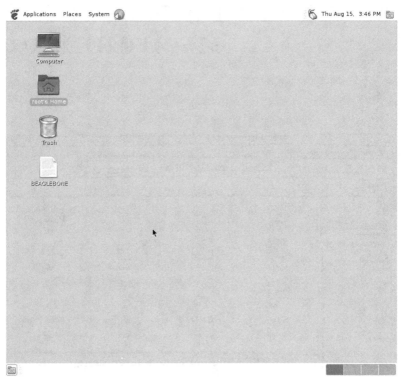

Figure 8-1. *The desktop environment on the BeagleBone.*

Accessing the Desktop

One way to access the desktop environment on the BeagleBone is to connect a keyboard, monitor, and mouse directly to it (see Figure 8-2). A USB keyboard and mouse can be connected through a USB hub to the USB host port on the board. If you have a BeagleBone Black, you can connect a monitor through the Micro HDMI port on the bottom of the board next to the MicroSD slot. This Micro HDMI connector, also called a "Type D" connector, isn't as easy to find in stores as the standard HDMI and the Mini HDMI type of cables. You may need to order this part online. Stores like Monoprice (*http://www.monoprice.com/*) sell inexpensive Micro HDMI cables and adapters.

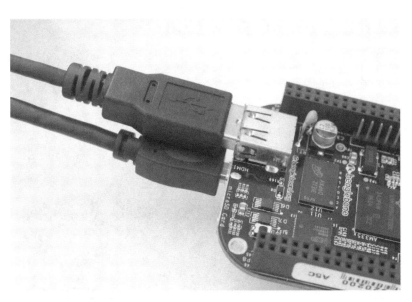

Figure 8-2. *Connecting a USB cable and micro HDMI cable to the Beagle-Bone Black*

 If you have the original BeagleBone, you can attach the DVI Cape expansion board to display the desktop environment on a monitor.

You can also access the desktop remotely through VNC if you'd rather not connect a keyboard, mouse, and monitor. See "Connecting to the Desktop Remotely with VNC" on page 106.

After connecting the monitor to your BeagleBone and rebooting the board, you should see the desktop environment on screen (Figure 8-1). If not, try clicking the mouse or hitting a key on the keyboard.

If you set a password for root ("Setting a Password" on page 29), you won't be logged into the desktop automatically and GNOME may get "stuck" trying to log you in. If you're stuck at the login screen and want to disable automatic login, first switch terminals ("Switching Terminals" on page 101) so that you can edit */etc/gdm/custom.conf* and change `TimedLoginEn able=true` to `TimedLoginEnable=false` and reboot the Beagle-Bone.

Getting to the Terminal

Despite the fact you can do a lot within the desktop environment, from time to time you'll find yourself needing a command line. Luckily, it's only a few clicks or keypresses away.

Using the Terminal Application

On the menu bar at the top of the screen, click Applications→System Tools→Terminal, which will bring up a windowed terminal. (See Figure 8-3.)

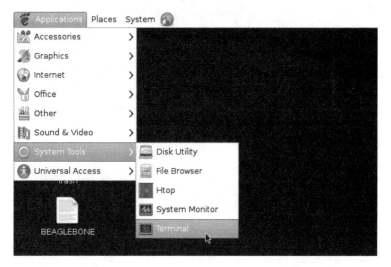

Figure 8-3. *Opening the Terminal application within GNOME*

You can have multiple windows open at once (File→Open Terminal) or keep multiple terminals open in a single window with a tabbed interface (File→Open Tab).

You may have noticed that the command prompt doesn't look like what you've gotten used to when you connect by SSH. It may appear as sh-4.2# instead of root@beaglebone:~#. To change that, click Edit→Preferences. In the Terminal Preferences window, select "Run command as login shell" (Figure 8-4) and close the preferences window. When you open a new terminal window or tab, you should see the same command prompt that you'd see when you log in via SSH.

When you open the terminal as a login shell, it automatically executes /etc/profile, a file which includes a few settings such as the style of command prompt, directories in the path (see "Executable Scripts" on page 52), and the default command line text editor.

Figure 8-4. *Changing Terminal's preferences*

Switching Terminals

You can also use the command line terminal without being within the desktop environment. To do this, try holding down the Control and Alt keys while tapping on the F1 through F6 keys. This gives you access to 5 different text-based terminals on F1 and F3 through F6. To get back to the desktop environment, type Control-Alt-F2.

Navigating the Filesystem

In Chapter 3, you explored the Linux filesystem from the command line with the commands cd and ls. To explore the filesystem within the desktop environment, double click on the Computer icon on the desktop. Then double click File System. The directories on display should look a bit familiar. You're viewing the contents of the root of the filesystem. (See Figure 8-5.)

You're exploring the filesystem as root, and you're using a filesystem viewer that lets you drag and drop. Be careful to not move anything important around.

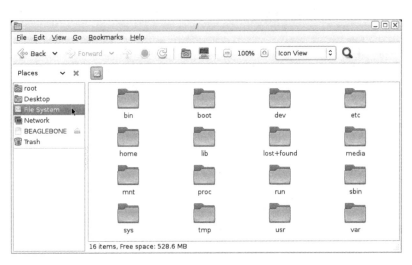

Figure 8-5. *The root of the filesystem as viewed through the desktop environment*

On the pane on the left side of the window are a few shortcuts to different parts of the filesystem. You can add to these shortcuts by dragging a file or folder to the pane. To remove a shortcut, right click on it and click Remove.

Most of the file management functionality that you'd expect from a modern operating system is available to you within GNOME. Open files and directories by double clicking on them, and take other actions by right-clicking on them. Moving files or directories is as simple as dragging them from window to window. If you want to copy instead of move, hold down the Control key before letting go of the file or directory.

Editing Text

A text editor called `gedit` is pre-installed on the BeagleBone. To launch it, click Applications→Accessories→gedit Text Editor. A window will open with a blank text document. If you're within the file management windows and want to open a file within `gedit`, simply right click on the file and click Open With→gedit Text Editor. (See Figure 8-6 for an example of opening a file with gedit and see Figure 8-7 for an example of its syntax highlighting.)

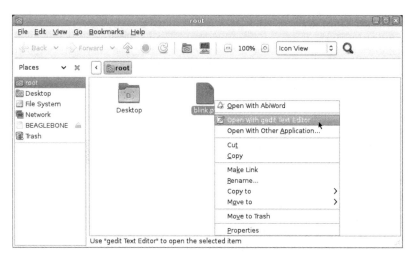

Figure 8-6. *Opening a file with* `gedit`

Figure 8-7. `gedit`'s *syntax highlighting makes some words bold or colored to make the code easier to read*

If you don't see gedit Text Editor in the menu after right clicking on a file, choose Open With Other Application...and select gedit from the list of applications. Make sure the option to remember this application is selected so that in the future you can just double click the file to open it.

`gedit` is also aware of the syntax for many programming languages and will use color to highlight keywords. For instance, if your file has the .py extension, `gedit` will recognize it as a Python file and color certain types of text appropriately.

Executing Scripts

At the beginning of many of the scripts we've written, we've included a *she-bang* or *hashbang* on the first line of the file. This instructs the system to use a particular interpreter (such as Python or Node) to run the file. That way, we can set the script file as executable and then run the file (see Figure 8-8). The system will then pass the script file to the interpreter listed on the line after the #!. The shebangs for the languages we've used are:

- Bash scripts: #!/bin/bash
- Python scripts: #!/usr/bin/python
- Node scripts: #!/usr/bin/node

Figure 8-8. *Setting a script as executable*

As long as the script has a proper shebang on the first line, you can set it as executable and run it directly. To set a file as executable within the desktop environment, right click on it and click Properties. Under the Permissions tab, check off the box that says "Allow executing file as program." This is the equivalent to executing `chmod +x filename` on the command line.

Now when you double click the file you'll be presented with the option to execute the script in terminal or display the file in a text editor. (See Figure 8-9.)

Figure 8-9. *After double clicking an executable script, you're given the option to run it or open the file in an editor*

When you choose Run in Terminal, you'll notice that the terminal window doesn't look as good and doesn't have the functionality as the one you launch from System Tools. Luckily, it's easy to change which terminal application launches in this case. On the menu bar, click System→Preferences→Preferred Applications and click on the System tab. Choose Xfce Terminal Emulator and close the window. Now when you double click executable scripts from file browser windows and choose Run in Terminal, it will launch in the nice Terminal window.

When you choose Display, the desktop environment will open the file in the default application for that file type. Sometimes it opens the file in AbiWord, a word processing application that's meant for creating and editing rich text documents. For editing code, it's much better to use a plain text editor. To change what application is launched when a particular type of file is opened for editing, right click on that type of file and click Properties. Under the Open With tab, select gedit Text Editor and click Close.

Switching Workspaces

If you find yourself with too many windows open and you're feeling overwhelmed, you can always switch to another workspace to get a clean desktop (see Figure 8-10). Just click on one of the inactive workspaces in the workspace switcher on the lower right hand side of the screen.

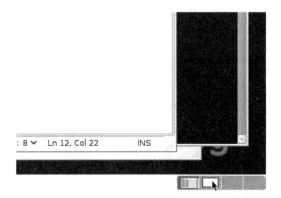

Figure 8-10. *Switching workspaces*

If four workspaces isn't enough or you'd like to label each workspace, you can do so in the preferences. Simply right click on the workspace switcher and click Preferences.

Connecting to the Desktop Remotely with VNC

You may want to access the BeagleBone's desktop environment but don't want to bother with connecting a keyboard, mouse, and monitor to the board. To access the desktop without these, you can use VNC, which is a way of displaying and controlling a computer's desktop environment from a remote computer over the network.

The first step is to start the VNC server on the BeagleBone. When you're logged in via SSH, run the following command, which sets the VNC to start when the destop environment starts.

```
root@beaglebone:~# opkg install angstrom-x11vnc-xinit
```

Then reboot your board so that the changes take effect.

If you're using a Mac, on the Finder's menu bar, click Go→Connect to Server. A window will appear where you can enter the server address. Enter `vnc://beaglebone.local:5900` (see Figure 8-11) and click connect. You'll be prompted to enter the password ("pass" in the example above) and then you'll be connected.

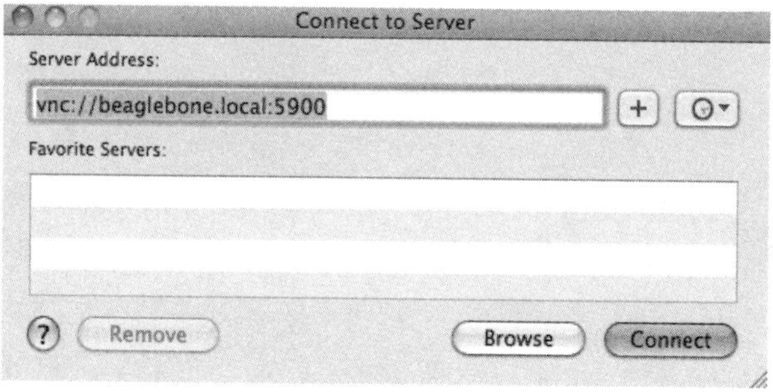

Figure 8-11. *Connecting to the BeagleBone via VNC on a Mac*

If you're using Windows, you'll need to download a VNC client application such as RealVNC (*http://realvnc.com/download/viewer/*) or TightVNC (*http://www.tightvnc.com/download.php*). After installing, connect to your BeagleBone with the server name `beaglebone.local:5900` or its IP address (Figure 8-12).

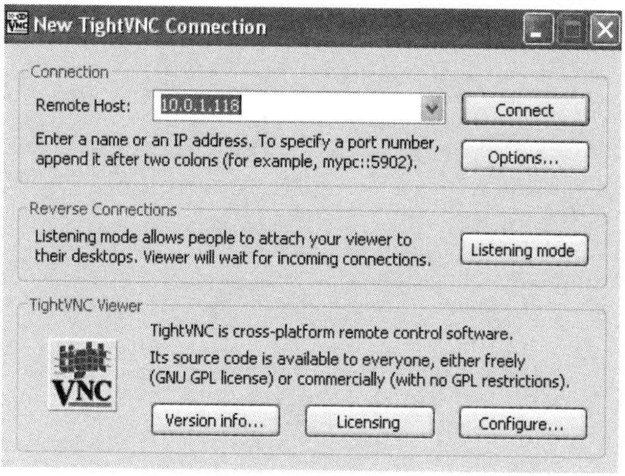

Figure 8-12. *Connecting to the BeagleBone via TightVNC on Windows*

9/Taking It Further

In this book, I've covered only the very basics when it comes to using the BeagleBone in physical computing projects. Hopefully I've equipped you with a few tools that can help make your projects a reality. Here are a few other resources I recommend for digging deeper:

- *bash Cookbook* by Carl Albing, JP Vossen, and Cameron Newham (O'Reilly)
- *Learning Python* by Mark Lutz (O'Reilly)
- *Make: Electronics* by Charles Platt (O'Reilly)

Getting Help

Let's face it, we've all run into trouble making projects. Sometimes, I don't even know where to start when it comes to turning a project idea into reality. Other than searching online for an answer, here are a few other resources that may come in handy while working with the BeagleBone:

BeagleBoard Google Group (*http://groups.google.com/group/beagleboard*)
   This group is quite active and covers both the BeagleBoard in addition to the BeagleBone.

BeagleBone Google Group (*http://groups.google.com/group/beaglebone*)
   Not as active as the BeagleBoard group, the group focuses on issues strictly related to the BeagleBone.

#beagle on Freenode IRC
   Internet chat rooms aren't dead! In fact, for getting a quick question answered, they can be the best resource. If you're unfamiliar with IRC, you can get acquainted at Freenode (*http://freenode.net/*). To jump right in, go to this Freenode page (*http://webchat.freenode.net/*), pick a nick, and join #beagle.

Getting Inspired

Many times, the inspiration for my projects comes from what other people are making and what technology they're using. Here are a few sites that post some of the best projects out there to help you find creative inspiration:

Makezine (*http://blog.makezine.com/*)
   Naturally, Makezine.com is an excellent place to find news and projects that makers of all types will be interested in.

Hack A Day (*http://hackaday.com/*)
   Make it a habit to read Hack A Day to get the scoop on some of the most advanced projects you'll find online.

Adafruit Blog (http://www.adafruit.com/blog/)
　　My friends at Adafruit have a knack for tracking down excellent projects that constantly serve as an inspiration for me.

Sharing Projects

At MAKE, we love to hear about all the projects you're working on, both the successes and failures. If you like to tell us about a project you're working on, you can submit it here (*http://blog.makezine.com/contribute/*). If you're looking for a way to full document your project build, post a tutorial to Make: Projects (*http://makeprojects.com*). Many projects in MAKE Magazine start off as Make: Projects!

To share with me directly, you can send me an email at *mattr@make zine.com* or you can find me on Twitter with the handle @MattRichardson.

Having Fun

There are many different reasons that people have for making things. For me, my desire to make things comes out of a passion for tinkering with technology. So many of my projects don't have a reason for existing other than the fact that making it was like scratching an itch. I didn't concern myself with answering the question "why make this?" It's like asking a skier why they go up a mountain only to ski down it.

It's also incredibly fun to go through the process of discovering what can be done with a new technology. Admittedly, figuring things out can get frustrating sometimes, but not much beats the feeling of mastering a new tool.

I encourage you to jump right in an mess around without the fear of failure. You just might surprise yourself.

A/Installing a Fresh Ångström Image

The BeagleBone comes pre-installed with Ångström, but if you want to reset your board with a clean install, sometimes your best bet is to download a new image. On the original BeagleBone, you'll be writing a bit-for-bit copy of an .img file to a MicroSD card that you boot off of when it's inserted into the board. You can do the same on the BeagleBone Black, but you also have the option of writing the disk image to the on-board flash memory, or eMMC, so that you can later boot without a MicroSD card inserted.

On OS X

1. Go to the BeagleBoard latest-images page (*http://beagleboard.org/latest-images*) and download the most recent Ångström demo image, which will be an .xz file.

   a. If you want to reflash the eMMC on the BeagleBone Black, choose the "eMMC Flasher" image. It contains the same operating system as the standard image, but is configured with a utility to rewrite that image from the MicroSD card to the eMMC.

   b. If you have the original BeagleBone or you want to boot from the MicroSD card on the BeagleBone Black, choose the standard image.

2. Download and install XZ Utils from the Mac OS X Packages page (*http://macpkg.sourceforge.net/*) so that you can uncompress the demo image.

3. In a terminal window, type **df**. This will list all the drive volumes connected to your computer.

4. Insert your MicroSD card into your computer and type **df** again.

5. The MicroSD card should now appear as new device on the list of disks. In my case, it's */dev/disk1*. Unmount that disk by typing

   ```
   sudo diskutil unmountDisk /dev/disk1
   ```

   Type in your computer's administrator password when sudo asks for it.

6. Navigate to the folder where you downloaded the Ångström demo image .xz file. For example:

```
cd ~/Downloads
```

7. Execute the following command to start the process of decompressing and copying the image to the card. Replace the name of the .xz file with the name of the file you downloaded and */dev/disk1* with the disk that you unmounted in the previous step:

```
sudo xz -dkc <Ångström Image File>.img.xz > /dev/disk1
```

 Make sure you are using the correct device filename, or you could overwrite the wrong disk, including the one your operating system is on!

8. The process of uncompressing and copying the image to the MicroSD can take up to an hour. You'll only see a blinking cursor while it's in progress, so just let it do its thing.

On Windows

1. Go to the BeagleBoard latest-images page (*http://beagleboard.org/latest-images*) and download the most recent Ångström demo image, which will be an .xz file.

   a. If you want to reflash the eMMC on the BeagleBone Black, choose the "eMMC Flasher" image. It contains the same operating system as the standard image, but is configured with a utility to rewrite that image from the MicroSD card to the eMMC.

   b. If you have the original BeagleBone or you want to boot from the MicroSD card on the BeagleBone Black, choose the standard image.

2. Go to the 7-zip main page (*http://www.7-zip.org/*) and download and install 7-Zip, which is what you'll use to decompress the .xz file.

3. Go to the Image Writer for Windows Launchpad page (*https://launchpad.net/win32-image-writer/*) and download and install Win32 Image Writer binary, which is what you'll use to write the disk image to the MicroSD card.

4. Navigate to the folder where you downloaded the Ångström demo image and right click on it. Under the 7-zip menu, click "Extract Files Here." This will create an .img file.

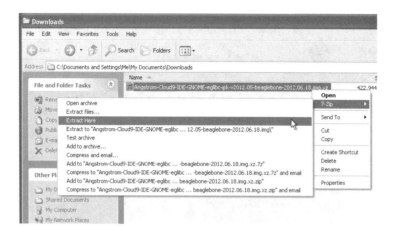

5. Insert your MicroSD card into your computer.

6. Open the .img in Win32 Image Writer and choose your MicroSD card as the device target.

7. Click "Write" to start writing the image to the card.

On Linux

1. Go to the BeagleBoard latest-images page (*http://beagleboard.org/latest-images*) and download the most recent Ångström demo image, which will be an .xz file.

   a. If you want to reflash the eMMC on the BeagleBone Black, choose the "eMMC Flasher" image. It contains the same operating system as the standard image, but is configured with a utility to rewrite that image from the MicroSD card to the eMMC.

   b. If you have the original BeagleBone or you want to boot from the MicroSD card on the BeagleBone Black, choose the standard image.

2. Get a listing of the storage devices connected to your computer:

   ```
   fdisk -l
   ```

3. Determine the device that matches your MicroSD card. It might be something like */dev/sda1*.

Make sure you are using the correct device filename, or you could overwrite the wrong disk, including the one your operating system is on!

4. Navigate to the folder where you downloaded the Ångström demo image .xz file. For example:

```
cd ~/Downloads
```

5. Execute the following command to start the process of decompressing and copying the image to the card. Replace the name of the .xz file with the name of the file you downloaded and */dev/sda1* with the device that matches your MicroSD card:

```
xz -dkc <Ångström Image File>.img.xz > /dev/sda1
```

6. The process of uncompressing and copying the image to the MicroSD can take up to an hour. You'll only see a blinking cursor while it's in progress, so just let it do its thing.

Flashing the eMMC

On the BeagleBone Black, you can download an image file for an *eMMC flasher*, which is a MicroSD card that is specifically configured to install the Ångström operating system from the MicroSD to the on-board flash memory so that you can boot without a MicroSD card inserted. If you used the steps above to create an eMMC flasher card, having it flash the BeagleBone Black's memory is easy:

1. With the power off on the BeagleBone Black, insert the eMMC flasher MicroSD card into the the MicroSD slot.

2. While holding the boot button (Figure A-1), connect the BeagleBone to power and keep holding the boot button for about fifteen seconds.

3. The USR LEDs will blink while the flashing process is taking place.

4. Flashing can take about 45 minutes. When it's done, all four USR LEDs will be lit solid.

5. Remove power from the board.

6. Remove the MicroSD card.

Figure A-1. *The boot button on the BeagleBone Black*

Now when you apply power to the board, it will boot off your newly flashed eMMC.

B/Setting up System Services

Throughout this book, you've launched projects by executing your code from the command line. However, there are many circumstances in which you'll want the BeagleBone to start your code immediately after it boots up. This is called a *system service*. For instance, in "Data Logging with Xively" on page 79, you created a utility to upload the current temperature to the Xively server every 20 seconds. After you launched it from the command line, it did just that until you exited the program, logged out, or powered down the Beagle-Bone.

Ångström uses `systemd` to manage all the system services. To change `systemd`'s settings, you'll use the `systemctl` command to enable, disable, start, stop, restart, and check the status of system services.

Creating a Service File

If you change into the */lib/systemd/system* directory and list its contents you'll see all the different *.service* files. These services aren't necessarily enabled, but many are. First you'll create a new *.service* file for your code in */lib/systemd/system*. I'll set the example from Example 6-11 as a system service in Example B-1.

```
root@beaglebone:/lib/systemd/system# nano xively-logger.service
```

In the new file you've created put in the following text from Example B-1:

Example B-1. Source code for /lib/systemd/system/ xively-logger.service

```
[Unit]
Description=Xively client ❶

[Service]
WorkingDirectory=/home/root/ ❷
ExecStart=/usr/bin/python xively-temp.py ❸
SyslogIdentifier=xively ❹
Restart=on-failure ❺
RestartSec=5 ❻
```

```
[Install]
WantedBy=multi-user.target ❼
```

❶ A human-readable description of the service.

❷ The location of the file you're launching.

❸ Command to execute within that location.

❹ Service name for system logs.

❺ If there's an error when trying to start the application, keep trying.

❻ Keep trying to restart the application every 5 seconds.

❼ When enabled, launch the service towards the end of the boot process (specifically, when the system is ready for multiple users to log into it).

Enabling and Starting the Service

After saving the file, enable the service with the command `systemctl enable`:

```
root@beaglebone:/lib/systemd/system# systemctl enable xively-logger
```

Now your service will start when the BeagleBone has booted up, but it's not running right now. To start it immediately, use the command `systemctl start`:

```
root@beaglebone:/lib/systemd/system# systemctl start xively-logger
```

If you want to restart the service, use `systemctl restart`:

```
root@beaglebone:/lib/systemd/system# systemctl restart xively-logger
```

Disabling and Stopping the Service

If you want to disable the service from starting up at boot time, you can run `systemctl disable`:

```
root@beaglebone:/lib/systemd/system# systemctl disable xively-logger
```

If the service was already running, disabling it won't stop it. Stopping the service is as easy as using `systemctl stop`:

```
root@beaglebone:/lib/systemd/system# systemctl stop xively-logger
```

Checking the Status of a Service

You can always check the status of a system service by using `systemctl status`:

```
root@beaglebone:/lib/systemd/system# systemctl status xively-logger
    Loaded: loaded (/lib/systemd/system/xively-logger.service; enabled)
    Active: active (running) since Wed 2013-07-24 14:57:03 EDT; 4s ago
  Main PID: 652 (python)
    CGroup: name=systemd:/system/xively-logger.service
            `-652 /usr/bin/python xively-temp.py

Jul 24 14:57:03 beaglebone systemd[1]: Starting Xively client...
Jul 24 14:57:03 beaglebone systemd[1]: Started Xively client.
```

You can also list enabled services by typing `systemctl` by itself:

```
root@beaglebone:/lib/systemd/system# systemctl
```

Type space to page through the list and q to go back to the command line.

Setting Time and Date as a System Service

After setting your timezone and testing the NTP servers in "Date and Time" on page 26, you may want to use `systemd` to have the BeagleBone set the time automatically on boot up. To do this, you'll have to edit `ntpdate.ser vice` in `/lib/systemd/system/`:

```
root@beaglebone:/lib/systemd/system# nano ntpdate.service
```

Modify the file so that it matches Example B-2.

Example B-2. Source code for /lib/systemd/system/ntpdate-logger.service

```
[Unit]
Description=Network Time Service (one-shot ntpdate mode)
Before=ntpd.service

[Service]
Type=oneshot
ExecStart=/usr/bin/ntpdate -b -s -u pool.ntp.org ❶
RemainAfterExit=yes
Restart=on-failure
RestartSec=5

[Install]
WantedBy=multi-user.target
```

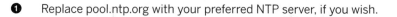 Replace pool.ntp.org with your preferred NTP server, if you wish.

At this point, the system service is set up and will work when you reboot your board. If you want to have your changes to the service take effect now, instruct `systemctl` to reload the configuration file:

```
root@beaglebone:/lib/systemd/system# systemctl --system daemon-reload
```

Then restart the service:

```
root@beaglebone:/lib/systemd/system# systemctl restart ntpdate
```

C/Quick Reference: GPIO

GPIO with the Filesystem

Exporting a pin

```
echo 44 > /sys/class/gpio/export
```

Setting a pin's direction to output

```
echo out > /sys/class/gpio/gpio44/direction
```

Writing a pin HIGH

```
echo 1 > /sys/class/gpio/gpio44/value
```

Writing a pin LOW

```
echo 0 > /sys/class/gpio/gpio44/value
```

Setting a pin's direction to input

```
echo in > /sys/class/gpio/gpio44/direction
```

Reading an input pin's value (will return 0 for low and 1 for high)

```
cat /sys/class/gpio/gpio44/value
```

GPIO with Python

Importing the Adafruit's BeagleBone IO Python Library

```
import Adafruit_BBIO.GPIO as GPIO
```

Setting pin's direction to output

```
GPIO.setup("P8_12", GPIO.OUT)
```

Writing a pin HIGH

```
GPIO.output("P8_12", GPIO.HIGH)
```

Writing a pin LOW

```
GPIO.output("P8_12", GPIO.LOW)
```

Setting a pin's direction to input

```
GPIO.setup("P8_14", GPIO.IN)
```

Reading an input pin's value (will return 0 or GPIO.LOW for low and 1 or GPIO.HIGH for high)

```
GPIO.input("P8_12")
```

Setting a pin for PWM (at 50% duty cycle)

```
import Adafruit_BBIO.PWM as PWM
PWM.start("P8_13", 50)
```

Changing the PWM duty cycle

```
PWM.set_duty_cycle("P8_13", 25)
```

Setting up analog input

```
import Adafruit_BBIO.ADC as ADC
ADC.setup()
```

Reading an analog input (returns a value between 0 and 1)

```
analogReading = ADC.read("P9_39")
```

GPIO with Node.js

Importing BoneScript

```
var b=require('bonescript');
```

Setting pin's direction to output

```
b.pinMode("P8_12", b.OUTPUT);
```

Writing a pin HIGH

```
b.digitalWrite("P8_12", b.HIGH);
```

Writing a pin LOW

```
b.digitalWrite("P8_12", b.LOW);
```

Setting a pin's direction to input

```
b.pinMode("P8_12", b.INPUT);
```

Reading an input pin's value (will return LOW or HIGH)

```
b.digitalRead("P8_12");
```

Setting a pin for PWM (at 50% duty cycle)

```
b.pinMode('P8_13', b.OUTPUT);
b.analogWrite('P8_13', 0.5);
```

Changing the PWM duty cycle

```
b.analogWrite('P8_13', 0.25);
```

Reading an analog input (returns a value between 0 and 1)

```
analogReading = b.analogRead('P9_39');
```

GPIO Pins

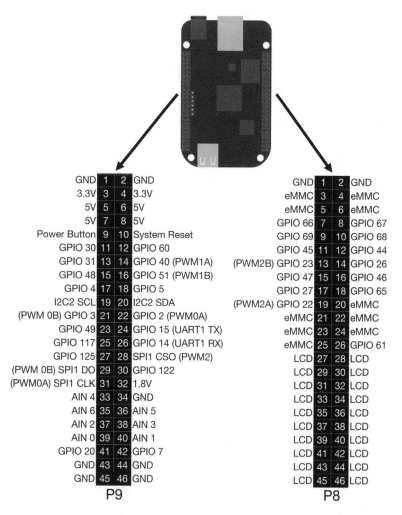

Figure C-1. *The default pin configuration on the BeagleBone (BeagleBone illustration courtesy of the Adafruit Fritzing library.)*

P9					P8			
GND	1	2	GND		GND	1	2	GND
3.3V	3	4	3.3V		eMMC	3	4	eMMC
5V	5	6	5V		eMMC	5	6	eMMC
5V	7	8	5V		GPIO 66	7	8	GPIO 67
Power Button	9	10	System Reset		GPIO 69	9	10	GPIO 68
GPIO 30	11	12	GPIO 60		GPIO 45	11	12	GPIO 44
GPIO 31	13	14	GPIO 40 (PWM1A)		(PWM2B) GPIO 23	13	14	GPIO 26
GPIO 48	15	16	GPIO 51 (PWM1B)		GPIO 47	15	16	GPIO 46
GPIO 4	17	18	GPIO 5		GPIO 27	17	18	GPIO 65
I2C2 SCL	19	20	I2C2 SDA		(PWM2A) GPIO 22	19	20	eMMC
(PWM 0B) GPIO 3	21	22	GPIO 2 (PWM0A)		eMMC	21	22	eMMC
GPIO 49	23	24	GPIO 15 (UART1 TX)		eMMC	23	24	eMMC
GPIO 117	25	26	GPIO 14 (UART1 RX)		eMMC	25	26	GPIO 61
GPIO 125	27	28	SPI1 CSO (PWM2)		LCD	27	28	LCD
(PWM 0B) SPI1 DO	29	30	GPIO 122		LCD	29	30	LCD
(PWM0A) SPI1 CLK	31	32	1.8V		LCD	31	32	LCD
AIN 4	33	34	GND		LCD	33	34	LCD
AIN 6	35	36	AIN 5		LCD	35	36	LCD
AIN 2	37	38	AIN 3		LCD	37	38	LCD
AIN 0	39	40	AIN 1		LCD	39	40	LCD
GPIO 20	41	42	GPIO 7		LCD	41	42	LCD
GND	43	44	GND		LCD	43	44	LCD
GND	45	46	GND		LCD	45	46	LCD

About the Author

Matt Richardson is a Brooklyn-based creative technologist and video producer. He's a contributor to *MAKE* magazine and *Makezine.com*. Matt is also the owner of Awesome Button Studios, a technology consultancy. Highlights from his work include the Descriptive Camera, a camera that outputs a text description of a scene instead of a photo. He also created The Enough Already, a DIY celebrity-silencing device. Matt's work has garnered attention from *The New York Times*, *Wired*, *New York Magazine* and has also been featured at The Nevada Museum of Art and at the Santorini Bienniele. He is currently a Master's candidate at New York University's Interactive Telecommunications Program.

The cover photo was created by Marc de Vinck. The cover and body font is BentonSans, the heading font is Serifa, and the code font is Bitstreams Vera Sans Mono.